An Economic Analysis on Automated Construction Safety

Rita Yi Man Li

An Economic Analysis on Automated Construction Safety

Internet of Things, Artificial Intelligence and 3D Printing

Springer

Rita Yi Man Li
Real Estate and Economics Research Lab,
Sustainable Real Estate Research Center,
Department of Economics and Finance
Hong Kong Shue Yan University
Hong Kong
Hong Kong

ISBN 978-981-10-5770-0 ISBN 978-981-10-5771-7 (eBook)
DOI 10.1007/978-981-10-5771-7

This book is supported by Hong Kong Shue Yan University RSDC grant.

Library of Congress Control Number: 2017946631

Printed on acid-free paper

This Springer imprint is published by Springer Nature
The registered company is Springer Nature Singapore Pte Ltd.
The registered company address is: 152 Beach Road, #21-01/04 Gateway East, Singapore 189721, Singapore

Contents

Chapter 1
Turning the Tide in the Construction Industry: From Traditional Construction Safety Measures to an Innovative Automated Approach

Abstract The construction industry has been viewed as labour intensive with many accidents occurring on sites around the world. Many construction companies have implemented various types of construction safety measures to reduce the likelihood of accidents on sites. We will first shed light on the conventional means to alleviate construction safety risks with an example of a large-scale company that rents a factory site to serve as a safety-training centre. Posters and slogans display at Seattle and Adelaide construction sites will illustrate more traditional forms of training and warnings. We then move on to provide a brief introduction to various kinds of automated construction tools, such as robots, virtual reality, the Internet of things, and additive manufacturing which completely transform traditional works in the construction industry. The objectives and research methods adopted in this book will also be stated.

Keywords Institutional economics · Cost and benefit · Automated construction safety

1 Introduction

The construction industry records greater fatal and nonfatal accident rates in comparison to many other industries around the world (Azhar and Choudhry 2016), at any time. In 2012, more than one in five fatal accidents at work occurred in the EU construction sector alone (Edirisinghe and Lingard 2016; Li and Poon 2011). In many circumstances, accidents on construction sites are not the results of an act of god but a series of human errors among various stakeholders together with other basket of factors (Table 1). Thus, some of the previous research concedes that the occurrence of the accidents are just the end results of a sequence of events (Li and Poon 2013a). In conclusion, the so-called once-in-a-blue-moon accidents not only adversely affect the construction industry's profit margins but also harm innovation strategies in the entire construction supply chain, the ability to deploy new technologies in the future and the best practices in the industry (Teizer 2016).

R.Y.M. Li, *An Economic Analysis on Automated Construction Safety*,
DOI 10.1007/978-981-10-5771-7_1

Table 1 Causes of construction accidents recorded in previous literatures from 2000 to 2017 [this table is a revised version of Li (2015a) and Li and Poon (2013d)]

Factors influencing accidents on sites	Literature
Workers	
Aged worker	Rameezdeen and Elmualim (2017)
Bricklayers and building labourers	Suárez-Cebador et al. (2014)
Fatigue	Chan (2011), Li and Ng (2017)
Human error	Garrett and Teizer (2009), Zhi et al. (2003)
Lack of or poor training	Chan et al. (2004), Debrah and Ofori (2001), Liu et al. (2007), Zahoor et al. (2017)
Lack of safety knowledge	Li (2006), Mitropoulos et al. (2005), Le et al. (2014), Li (2015a)
Migrant	Debrah and Ofori (2001), Hassan and Houdmont (2014)
Poor materials handling	Irumba (2014)
Poor safety attitude	Toole (2002), Teo et al. (2005), Yu et al. (2014)
Stress	Irumba (2014)
Workers' actions, behaviours, capabilities and characteristics	Gibb et al. (2014), Khosravi et al. (2014), Li et al. (2015b), Dzeng et al. (2016)
Young	Li (2006), Chi et al. (2005)
Management	
Communication	Motter and Santos (2017)
Construction task planning	Akhmad et al. (2001)
Design	Gambatese et al. (2008), Arocena and Núñez (2010), Kongtip et al. (2008), Bong et al. (2015), Malekitabar et al. (2016)
Housekeeping	Toole (2002), Haslam et al. (2005a), Hu et al. (2011), Ahmad et al. (2016)
Protective equipment and equipment for work	Toole (2002), Haslam et al. (2005b), Eliufoo (2007), Cheng and Wu (2013), Chong and Low (2014), Gibb et al. (2014), Ahmad et al. (2016)
Project management in general	Jabbari and Ghorbani (2016), Khosravi et al. (2014), Lingard et al. (2017)
Relationship with the crew	Debrah and Ofori (2001)
Safety climate and culture	Li (2015a), Ling et al. (2009), Goh et al. (2016), He et al. (2016)
Size of the companies	Lin and Mills (2001)
Subcontract	Debrah and Ofori (2001), Toole (2002)
Traditional construction methods	Chun et al. (2012)
Weather	
Hot Summer	Navon and Kolton (2006), Chan (2011), Hu et al. (2011)
Poor visibility	Arditi et al. (2005)
Working environment	
Hectic schedule	Debrah and Ofori (2001)

(continued)

Table 1 (continued)

Factors influencing accidents on sites	Literature
Heights below 30 feet	Cakan et al. (2014), Huang and Jimmie (2003)
Site layout	Gibb et al. (2014)
Small alteration projects	Cakan et al. (2014)
Structural failure	Hintikka (2011)
Complex work or unsafe working condition	Choi et al. (2011), Chockalingam and Sornakumar (2011), Shin et al. (2014a)
Technical failure	Raviv et al. (2017a)
Types of work	Grant and Hinze (2014)
Multi-storied and large-sized construction	Shin et al. (2014b)
Legal system	
Regulations and enforcement	Chockalingam and Sornakumar (2011), Chan et al. (2004), Zahoor et al. (2017)
Time	
Afternoon	Gurcanli and Mungen (2013), Ahmad et al. (2016)
Economic	
Piece rate	Debrah and Ofori (2001)
Projects of low construction cost	Huang and Jimmie (2003)
Low spending on safety issues	Debrah and Ofori (2001)
Firms' profitability increases	Forteza et al. (2017)

As a matter of fact, the consistently high accident rates not only lead to insurmountable compensation costs but also a great amount of non-monetary loss. Safety officers and practitioners explore many different means to reduce the costs of accidents, often beyond what is required by the established regulations. Nevertheless, many countries' construction practitioners may have realised that the so-called best safety practices in the industry have already reached a plateau. In view of this, innovative approaches are necessary for a further reduction in construction incidents (Saurin et al. 2005).

2 Traditional Construction Safety Measures

Traditionally, construction companies provide safety training to workers through face-to-face lectures, "tool box talks" and learning activities (Li and Poon 2013a; Li 2015a). For example, a particular large-scale construction company rents a factory unit in Hong Kong containing all types of safety training equipment. Figure 1 illustrates the safety belt training where the instructor is demonstrating how to use the safety belt in the correct manner. Figures 2 and 3 illustrate protective equipment, such as safety helmets and gloves. Figures 4 and 5 show two major works with higher hazard levels: work at height and excavation work. Figures 6 and 7 display various types of anchors which are used on sites. Figures 8, 9, 10, 11 and 12

Fig. 1 Safety belt training (author's photo)

Fig. 2 Different types of personal protective equipment displayed, such as helmet and shoes (*left*) (author's photo)

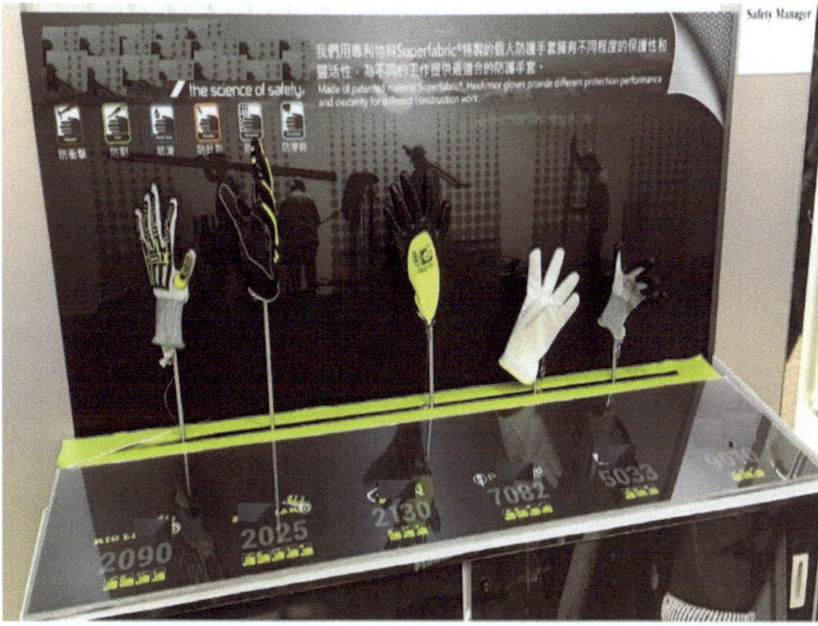

Fig. 3 Safety gloves for different types of work (*right*) (author's photo)

Fig. 4 Working at height model (*left*) (author's photo)

Fig. 5 A model reminding workers to operate safely during excavation work (*right*) (author's photos)

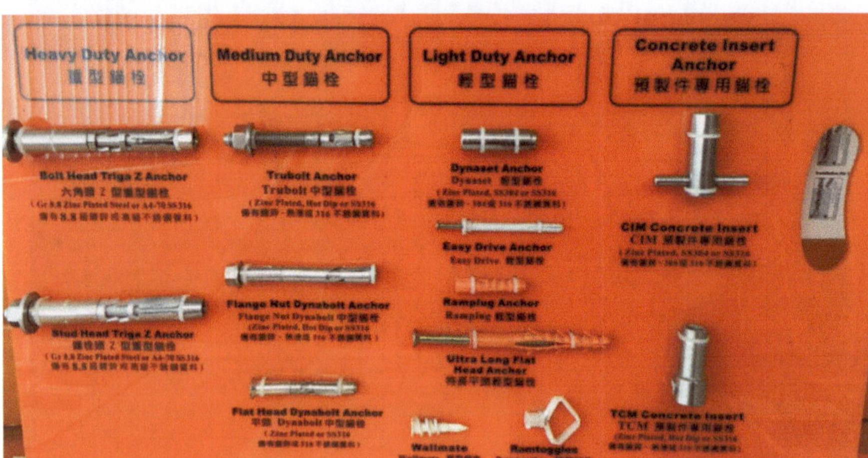

Fig. 6 Different types of anchors for precast concrete (author's photo)

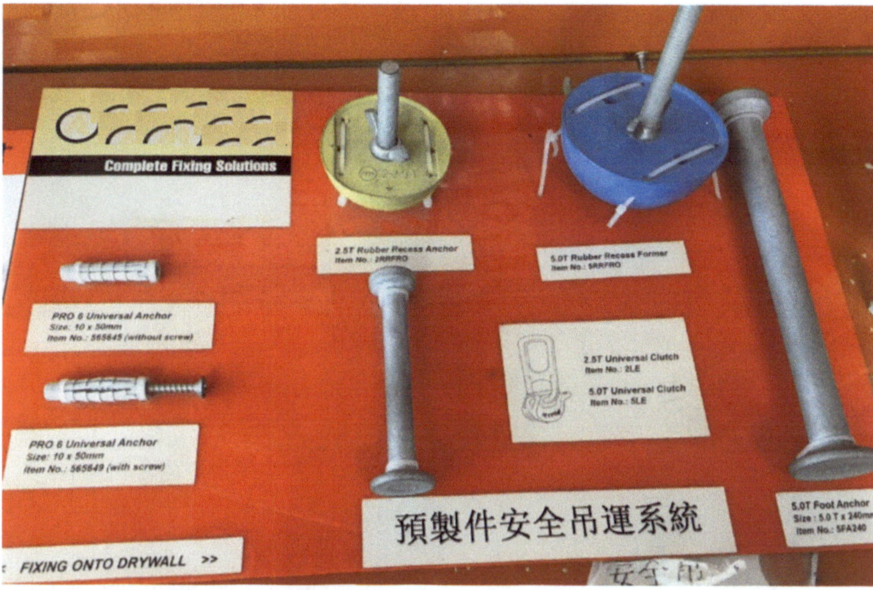

Fig. 7 Different types of anchor used for transporting materials (author's photo)

Fig. 8 Construction site in Seattle warning trespassers against entering the site and causing accidents (author's photo)

Fig. 9 "Please go home safely tonight" slogan, placed at the entrance of a construction site in Adelaide (author's photo)

elucidate posters display on sites in different parts of the World and Fig. 13 demonstrates the eye-catching orange clothing for road repair works in Seattle.

Nevertheless, in recent years, technological breakthroughs in various types of information technology have provided a golden opportunity to improve safety on sites via some innovative approaches. For example, this company has incorporated virtual reality training in this learning centre, on which more information will be included in a later chapter. It has also adopted various different kinds of automated tools, such as additive manufacturing for three-dimensional model printing.

Fig. 10 A poster is placed at the entrance to ensure all personnel must be inducted prior to working on-site and that no unauthorised access occurs at the Royal Adelaide Hospital construction site (author's photo)

Fig. 11 Poster advocating personal protective equipment at the Royal Adelaide Hospital construction site (author's photo)

Fig. 12 Safety and construction site notice board near the entrance of the construction site (author's photo)

Fig. 13 Workers wear eye-catching orange clothing for road repair works in Seattle (author's photo)

3 Why Is an Automated Construction Process Necessary? A Bird's Eye View of Recent Automated Construction Technologies' Development

The need to enhance quality, improve productivity and reduce costs raises the pace of automation in the construction industry (Shehab 2009). For example, radio-frequency identification (RFID) systems are employed for tracking construction assets, laser scanning, web-based applications areas or/and a hybrid system of two or more technologies to monitor the progress of various departments on construction sites. On the other hand, in view of the prohibitively high costs of accidents, the means for managing site safety has always been a high priority necessity. The advancement in sensing, wearable robotics and the Internet of things (IoT) has not only redeveloped the entire outlook of the construction industry, but may be considered as the greatest advancement in the industry for many years (Teizer 2016).

3.1 Building Information Modelling

Built upon the concepts of three-dimensional modelling by incorporating non-graphical object data into the model, BIM generally refers to a modelling technology with a set of processes to produce, analyse and communicate building models. BIM is also defined as comprehensive information accumulation with regard to the design, construction and building operation, anchored to a geometric two- or three-dimensional model of the intended building (Demian and Walters 2014).

Previous studies show that BIM's benefits include parametric modelling and detailed building analysis (Demian and Walters 2014), data omission minimization (Park and Cai 2017), time and costs reduction (Ciribini et al. 2016). Whilst BIM generally refers to three-dimensional X–Y–Z modelling, there is also four-dimensional BIM, whereby the timeline of the construction programme is linked into the three-dimensional building model. (Li 2017, forthcoming) On the other hand, Bansal (2011) suggests that GIS can be used togethe with BIM, such that both 3D model is linked with its surrounding topography as 4D BIM. Five-dimensional models include cost data in addition to the four-dimensional modelling (Demian and Walters 2014). Six-dimensional BIM includes facility management, such that the warranty, locations of ducts and conduits are included.[1]

Vysotskiy et al. (2015) mentioned that the global trend in using BIM technology involved three-dimensional design and life cycle analysis. In the 2012 London Olympics, BIM combined various data sources and monitored the construction works on sites. Some popular BIM software, such as AutoCAD and Autodesk, can reduce errors by between 50 and 90%. More importantly, safety and risk management information can be added to BIM which aims to reduce the likelihood of construction accident on sites (Ding et al. 2016; Ganah and John 2015; Kim et al. 2016; Park and Kim 2013; Zhang et al. 2013).

3.2 Additive Manufacturing

Additive manufacturing creates three-dimensional solid models from digital designs (Wang et al. 2014b), architectural modelling (Buswell et al. 2007) to print housing on earth (Wu et al. 2016) and building on the moon (Cesaretti et al. 2014). Materials for 3D printing vary and depend on the needs of the construction works. For example, Henke and Treml (2013) use wood based bulk material and cement as binder for additive manufacturing in construction works. Kazemian et al. (2017) and Lim et al. (2012) propose different approaches to print concrete. Academic researchers are optimistic that this advancement may reduce the number of labourers on sites, construction costs and time, whilst also increasing architectural freedom, allowing for sophisticated designs for aesthetic and structural purposes[2] (Xia and Sanjayan 2016; Bassoli et al. 2007).

---

[1]Author's interview results with two construction companies in Hong Kong.

[2]Although the authors suggest safety as one of the merits, they have not elaborated on the relationship between 3D-printing and construction safety.

3.3 Virtual Reality (VR)

VR creates an interactive computer-simulated real-world environment, which produces the sensation of the user actually being in situ (Freeman et al. 2017). Virtual reality has been used to visualise of details of the bridge component during construction (Sampaio and Martins 2014) and train and pomote workers in construction safety issues (Gammon Construction 2016; Sacks et al. 2013; Zhao and Lucas 2015). Safety information in visual form ensures that all important information is communicated to workers who may have low levels of literacy (Edirisinghe and Lingard 2016).

3.4 Internet of Things (IoT)

Internet of Things refers to unambiguously identifiable things that collect and exchange data amongst themselves and humans via computer networks and the Internet (Podgórski et al. 2017). In the construction industry, workers location tracking concepts are extended to monitor ergonomics and productivity. Remote sensing technology, for example, is used to record and analyse the precise position of the workforce, materials and equipment to enhance safety on sites (Teizer 2016). The trajectories of workers and crane load information were collected via a laser scanner. The results were recorded in real time and visualised in a three-dimensional range point cloud. Preliminary semi-automated trajectory analysis was then conducted when the workers worked in the hazardous excavated pit areas, such as confined or restricted spaces, when and where they were identified (Teizer 2016).

Apart from recording the positions of workers and materials, IoT is also used as safety warning system. It is imbued with devices that solicit, analyse and share safety information, details construction participants and offers an interconnected sensor monitoring network. It is a distributed and dynamic network with the capability to create an intelligent loop of safety checking, forecast and control. The IoT-based early warning safety system integrates smart sensor technology, such as piezoelectric and/or FBG sensors with location tracking technology, including WSN and/or RFID. The system monitors construction workers and all aspects on-site, sharing real-time safety information during construction (Ding et al. 2013).

3.5 Robots

Human error accounts for between 80–90% of on-site construction accidents (Raviv et al. 2017b). Monitoring dangerous work on construction sites can reduce the likelihood of accidents amongst manual labourers. Furthermore, well-designed robots can increase productivity on sites. Faster works can improve overall construction progress on sites. Besides, robots enables new and innovative methods for construction, architecture design, and implementation (Bloss 2014).

In South Korea, robots assemble steel beams and transport the bolting devices to the target bolting positions, overpassing the H-beam from one position to another and landing the RBA system on the floor (Jung et al. 2013). It is also used for laying brick (Yu et al. 2009). In Michigan, robots are used to autonomously identify, grasp and assemble prismatic building components with the help of MATLAB Calibration Toolbox algorithms, a visual marker-based metrology to establish local reference frames and detect the staged building components (Feng et al. 2015). Initiated by a group of ETH Zurich researchers, robots' architectural morphologies allow for the addressing of both functional and visually appealing concerns (Willmann et al. 2016). In Hong Kong, robots have recently been used to instal large-scale window panels (Gammon Construction 2016).

3.6 *Software Engineering for Construction Safety*

A software engineer must first identify the safety requirements at the system level and then ensure that the software meets the requirements. Formal verification and testing are also used to verify the software's functional correctness. Nevertheless, software correctness cannot ensure the safe operation of safety-critical software systems. To ensure that potentially hazardous causes cannot occur, the software must be verified against its safety requirements that are identified in safety analysis (Abdulkhaleq et al. 2015).

In short, the above-mentioned technologies not only change the traditional stance that working on-site is a human-centred industry, but also present important implications on construction safety, in the main. Nevertheless, despite the previous strands of the literature concerning top-of-the-range technologies, there are few or no research studies on the carrot-and-stick approach in adopting these applications (AM, VR, IoT, robots) from economic perspectives, such as the costs, benefits and institutional economics. One of the aims of this research is to fill this research gap.

4 Major Hurdles in Moving from Manual Work to an Automated Construction Approach

4.1 *Economic Costs*

More often than not, economic factors (either from the CBA or institutional perspectives) affect contractors and clients' decisions in adopting new technologies across the board. No firms dare try new technology if the cost of the tools is prohibitive, with limited benefits. Automated tools designed based on pure blue-sky research without consideration on economic aspect, however, will usually go belly up.

4.2 Institutions and Technological Change

As a matter of fact, technological changes and social institutions are largely interdependent (Bathelt and Glückler 2013). Informal institutions manifest themselves through social interactions and patterns of behaviour, including subordination, trust or collaboration, where formal institutions target mostly legal issues (Krammer 2015). It is often observed that institutional rather than functional change provides a vivid explanation for technological change. Even though established institutions are more likely to be associated with pre-existing structures and stability, new innovations require substantial alterations of the institutions. The interrelations with other institutions often prohibit technological change (Bathelt and Glückler 2013). Besides, economic institutions co-evolve with technologies due to informal cultural institutions, leaving considerable doubt concerning how new institutions exist without fatal resistance (Leonard and Granville 2010). Furthermore, top managers' resistance to change and excess formalisation often lead to organisational inflexibility in a dynamic environment (Fuentelsaz et al. 2015).

Economic agents may engage with different rationalities such as (1) recursive rationality where agents try to anticipate changes and shape the environment actively; (2) procedural rationality, which breaks down the problems and solves them step-by-step; or (3) instrumentalist rationality which mainly sheds light on reactive problem-solving in a stable environment. Some of the institutions become a burden which limits the opportunities of economic agents, leading to failure to search for the best practice solutions (Bathelt and Glückler 2013). Another dimension of institution (knowledge-related institutions which are, in turn, related to education and training) (Figueiredo 2016) also affects the likelihood of adopting automated tools on site.

5 Should We Adopt the New Technologies? A Cost Benefit Analysis (CBA) Approach

Previously, we highlighted the benefit that new technology can offer by using the automated tools on site: next, we shall view the adoption of these tools via the CBA analysis, recognised as one of the most widely accepted instruments in empirical research analysis due to its rational and systematic decision-making support tool (Djukic et al. 2016). Benefits of CBA include the generation and sharing of new knowledge, the transfer of related technological knowledge that passes to other firms in the supply chain, human capital formation via education and training and the cultural impact of the project. Costs include energy, maintenance, labour, communication, materials and negative externalities, such as noise or air pollution (Battistoni et al. 2016, forthcoming). Cost–benefit exercises analyse the necessity of strategies. If a project's private benefits exceed the private costs, the project will be implemented (Scott 2009).

6 Objectives, Hypothesis and Research Methods

In this book, the objectives are twofold

1. To provide an up-to-date specifications with regard to the recent development and adoption of various automated tools, such as building information modelling (BIM), additive manufacturing (AM), virtual reality (VR), the Internet of things (IoT), robotics' application on sites and their impact on worldwide construction safety;
2. To analyse the costs and benefits of the automated tools, with regard to international construction safety, by revealing the opinions of construction practitioners, academics and tool providers, in order to gain more knowledge about the agents that encourage our practitioners to use the automated tools.

7 Conclusion

Construction accidents happen on sites due to a basket of factors. Traditionally, construction accidents are prevented through training. For example, a particularly large company in Hong Kong rent a factory for training their staff. Various protective equipments are displayed and workers given chance to try to use various tools before they use them on sites. Posters are posted on sites to remind the workers to take all the safety precautions, workers have to wear the eye-catching clothes to avoid traffic accidents when they work for road repairmen works and so on. To avoid any trespassers enter construction sites, posters are posted outside all the construction sites in Seattle to warn trespassers against entering the site.

In recent years, technological breakthrough has changed the façade of the construction industry. Various automated tools, such as BIM, AM, VR, IoT, robotics and so on are applied on sites. Nevertheless, studies with regard to their impact on construction safety are quite scarce. This book aims to fill this research gap from cost and benefits perspectives. Academics, tool providers and construction practitioners will be included in the interviews and RAND appropriateness study.

References

Abdulkhaleq, A., S. Wagner, and N. Leveson. 2015. A comprehensive safety engineering approach for software-intensive systems based on stpa. *Procedia Engineering* 128: 2–11.

Ahmad, S., M. Iraj, M. Abbas, and A. Mahdi. 2016. Analysis of occupational accidents induced human injuries: A case study in construction industries and sites. *Journal of Civil Engineering and Construction Technology* 7: 1–7.

Akhmad, S., A.R. Duff, and J.P. Stephen. 2001. Development of causal model of construction accident causation. *Journal of Construction Engineering and Management* 127: 337–344.

Arditi, D., M. Ayrancioglu, and J.J. Shi. 2005. Worker safety issues in night-time highway construction. *Engineering, Construction and Architectural Management* 12: 487–501.

Arocena, P., and I. Núñez. 2010. An empirical analysis of the effectiveness of occupational health and safety management systems in smes. *International Small Business Journal* 28: 398–419.

Azhar, S., and R.M. Choudhry. 2016. Capacity building in construction health and safety research, education, and practice in Pakistan. *Built Environment Project and Asset Management* 6: 92–105.

Bansal, V.K. 2011. Application of geographic information systems in construction safety planning. *International Journal of Project Management* 29: 66–77.

Bassoli, E., A. Gatto, L. Iuliano, and M.G. Violante. 2007. 3D printing technique applied to rapid casting. *Rapid Prototyping Journal* 13: 148–155.

Bathelt, H., and J. Glückler. 2013. Institutional change in economic geography. *Progress in Human Geography* 38: 340–363.

Battistoni, G., M. Genco, M. Marsilio, C. Pancotti, S. Rossi, S. Vignetti. 2016, forthcoming. Cost-benefit analysis of applied research infrastructure. Evidence from health care. *Technological Forecasting and Social Change.*

Bloss, R. 2014. Robots have come to architecture to model, construct, fabricate and offer new approaches to create innovative designs, elements and structures. *Industrial Robot: An International Journal* 41: 403–407.

Bong, S., R. Rameezdeen, J. Zuo, R.Y.M. Li, and G. Ye. 2015. The designer's role in workplace health and safety in construction industry: Post-harmonized regulations in South Australia. *International Journal of Construction Management* 15: 276–287.

Buswell, R.A., R.C. Soar, A.G.F. Gibb, and A. Thorpe. 2007. Freeform construction: Mega-scale rapid manufacturing for construction. *Automation in Construction* 16: 224–231.

Cakan, H., E. Kazan, and M. Usmen. 2014. Investigation of factors contributing to fatal and nonfatal roofer fall accidents. *International Journal of Construction Education and Research* 10: 300–317.

Cesaretti, G., E. Dini, X. De Kestelier, V. Colla, and L. Pambaguian. 2014. Building components for an outpost on the lunar soil by means of a novel 3D printing technology. *Acta Astronautica* 93: 430–450.

Chan, A.H.S., W.Y. Kwok, and V.G. Duffy. 2004. Using AHP for determining priority in a safety management system. *Industrial Management & Data Systems* 104: 430–445.

Chan, M. 2011. Fatigue: The most critical accident risk in oil and gas construction. *Construction Management and Economics* 29: 341–353.

Cheng, C.W., and T.C. Wu. 2013. An investigation and analysis of major accidents involving foreign workers in Taiwan's manufacture and construction industries. *Safety Science* 57: 223–235.

Chi, C.F., T.C. Chang, and H.-I. Ting. 2005. Accident patterns and prevention measures for fatal occupational falls in the construction industry. *Applied Ergonomics* 36: 391–400.

Chockalingam S, and Sornakumar. 2011. Tools for improving safety performance of indian construction industry-AWH & SIT approach. European Journal of Economics, Finance and Administrative Sciences 35: 15–22.

Choi, T.N.Y., D.W.M. Chan, and A.P.C. Chan. 2011. Perceived benefits of applying pay for safety scheme (PFSS) in construction—a factor analysis approach. *Safety Science* 49: 813–823.

Chong, H.Y., and T.S. Low. 2014. Accidents in Malaysian construction industry: Statistical data and court cases. *International Journal of Occupational Safety and Ergonomics* 20: 503–513.

Chun, C.K., H. Li, and M. Skitmore. 2012. The use of virtual prototyping for hazard identification in the early design stage. *Construction Innovation: Information, Process, Management* 12: 29–42.

Ciribini, ALC, S. Mastrolembo Ventura, M. Paneroni. 2016. Implementation of an interoperable process to optimise design and construction phases of a residential building: A BIM pilot project. Automation in Construction. 71: 62–73.

Debrah, Y.A., and G. Ofori. 2001. Subcontracting, foreign workers and job safety in the Singapore construction industry. *Asia Pacific Business Review* 8: 145–166.

Demian, P., and D. Walters. 2014. The advantages of information management through building information modelling. *Construction Management and Economics* 32: 1153–1165.
Ding, L.Y., C. Zhou, Q.X. Deng, H.B. Luo, X.W. Ye, Y.Q. Ni, et al. 2013. Real-time safety early warning system for cross passage construction in yangtze riverbed metro tunnel based on the internet of things. Automation in Construction 36.
Ding, L.Y., B.T. Zhong, S. Wu, and H.B. Luo. 2016. Construction risk knowledge management in BIM using ontology and semantic web technology. *Safety Science* 87: 202–213.
Djukic, M., I. Jovanoski, O.M. Ivanovic, M. Lazic, and D. Bodroza. 2016. Cost-benefit analysis of an infrastructure project and a cost-reflective tariff: A case study for investment in wastewater treatment plant in Serbia. *Renewable and Sustainable Energy Reviews* 59: 1419–1425.
Dzeng, R.-J., C.-T. Lin, and Y.-C. Fang. 2016. Using eye-tracker to compare search patterns between experienced and novice workers for site hazard identification. *Safety Science* 82: 56–67.
Edirisinghe, R., and H. Lingard. 2016. Exploring the potential for the use of video to communicate safety information to construction workers: Case studies of organizational use. *Construction Management and Economics* 34: 366–376.
Eliufoo, H.K. 2007. Gendered division of labour in construction sites in Zanzibar. *Women in Management Review* 22: 112–121.
Feng, C., Y. Xiao, A. Willette, W. McGee, and V.R. Kamat. 2015. Vision guided autonomous robotic assembly and as-built scanning on unstructured construction sites. *Automation in Construction* 59: 128–138.
Figueiredo, P.N. 2016. Evolution of the short-fiber technological trajectory in brazil's pulp and paper industry: The role of firm-level innovative capability-building and indigenous institutions. *Forest Policy and Economics* 64: 1–14.
Forteza, F.J., J.M. Carretero-Gómez, and A. Sesé. 2017. Occupational risks, accidents on sites and economic performance of construction firms. *Safety Science* 94: 61–76.
Freeman, D., S. Reeve, A. Robinson, A. Ehlers, D. Clark, B. Spanlang, et al. 2017. Virtual reality in the assessment, understanding, and treatment of mental health disorders. Psychological Medicine 1–8.
Fuentelsaz, L., E. Garrido, and J.P. Maicas. 2015. Incumbents, technological change and institutions: How the value of complementary resources varies across markets. *Strategic Management Journal* 36: 1778–1801.
Gambatese, J.A., M. Behm, and S. Rajendran. 2008. Design's role in construction accident causality and prevention: Perspectives from an expert panel. *Safety Science* 46: 675–691.
Gammon Construction. 2016. Gammon construction extends innovative technology efforts to optimize construction—virtual reality safety training, GPS tracking, robotic construction. Available: http://www.gammonconstruction.com/en/html/press/press-6b9ebe23ad3a426b847f03c4e5fa7031.html. Accessed 7 Aug 2016.
Ganah, A., and G.A. John. 2015. Integrating building information modeling and health and safety for onsite construction. *Safety and Health at Work* 6: 39–45.
Garrett, J.W., and J. Teizer. 2009. Human factors analysis classification system relating to human error awareness taxonomy in construction safety. *Journal of Construction Engineering & Managemen* 135: 754–763.
Gibb, A., H. Lingard, M. Behm, and T. Cooke. 2014. Construction accident causality: Learning from different countries and differing consequences. *Construction Management and Economics* 32: 446–459.
Goh, Y.M., and M.J. Askar Ali. 2016. A hybrid simulation approach for integrating safety behavior into construction planning: An earthmoving case study. *Accident Analysis and Prevention* 93: 310–318.
Grant, A., and J. Hinze. 2014. Construction worker fatalities related to trusses: An analysis of the OSHA fatality and catastrophic incident database. *Safety Science* 65: 54–62.
Gurcanli, G.E., and U. Mungen. 2013. Analysis of construction accidents in Turkey and responsible parties. *Industrial Health* 51: 581–595.

Haslam RA, Hide SA, Gibb AGF, Gyi DE, Pavitt T, Atkinson S, et al. 2005a. Contributing factors in construction accidents Applied Ergonomics 36:401–415.

Haslam, R. A., S. A. Hide, G. F.Gibb, D. E. Gyi, S. Atkinson, and A. R. Duff. 2005b. Contributing factors in construction accidents. Applied Ergonomics 36:401–415.

Hassan, H.A., and J. Houdmont. 2014. Health and safety implications of recruitment payments in migrant construction workers. *Occupational Medicine* 64: 331–336.

He, Q., S. Dong, T. Rose, H. Li, Q. Yin, and D. Cao. 2016. Systematic impact of institutional pressures on safety climate in the construction industry. *Accident Analysis and Prevention* 93: 230–239.

Henke, K., and S. Treml. 2013. *European Journal of Wood and Wood Products* 71: 139–141

Hill, J., W. Randolph Ford, and I.G. Farreras. 2015. Real conversations with artificial intelligence: A comparison between human–human online conversations and human–chatbot conversations. *Computers in Human Behavior* 49: 245–250.

Hintikka, N. 2011. Accidents at work during temporary agency work in Finland—comparisons between certain major industries and other industries. *Safety Science* 49: 473–483.

Hu, K., H. Rahmandad, T. Smith-Jackson, and Woodrow Winchester. 2011. Factors influencing the risk of falls in the construction industry: A review of the evidence. *Construction Management and Economics* 29: 397–416.

Huang, X., and H. Jimmie. 2003. Analysis of construction worker fall accidents. *Journal of Construction Engineering and Management* 129: 262–271.

Irumba, R. 2014. Spatial analysis of construction accidents in Kampala, Uganda. *Safety Science* 64: 109–120.

Jabbari, M., and R. Ghorbani. 2016. Developing techniques for cause-responsibility analysis of occupational accidents. *Accident Analysis and Prevention* 96: 101–107.

Jung, K., B. Chu, and D. Hong. 2013. Robot-based construction automation: An application to steel beam assembly (part ii). *Automation in Construction* 32: 62–79.

Kazemian, A., X. Yuan, E. Cochran, and B. Khoshnevis. 2017. Cementitious materials for construction-scale 3d printing: Laboratory testing of fresh printing mixture. *Construction and Building Materials* 145: 639–647.

Khosravi, Y., H. Asilian-Mahabadi, E. Hajizadeh, N. Hassanzadeh-Rangi, H. Bastani, and A.H. Behzadan. 2014. Factors influencing unsafe behaviors and accidents on construction sites: A review. *International Journal of Occupational Safety and Ergonomics* 20: 111–125.

Kim, K., Y. Cho, and S. Zhang. 2016. Integrating work sequences and temporary structures into safety planning: Automated scaffolding-related safety hazard identification and prevention in BIM. *Automation in Construction* 70: 128–142.

Kongtip, P., W. Yoosook, and S. Chantanakul. 2008. Occupational health and safety management in small and medium-sized enterprises: An overview of the situation in Thailand. *Safety Science* 46: 1356–1368.

Krammer, S.M.S. 2015. Do good institutions enhance the effect of technological spillovers on productivity? Comparative evidence from developed and transition economies. *Technological Forecasting and Social Change* 94: 133–154.

Le, Q.T., D.Y. Lee, and C.S. Park. 2014. A social network system for sharing construction safety and health knowledge. *Automation in Construction* 46: 30–37.

Leonard, C.S., and B. Granville. 2010. Do informal institutions matter for technological change in Russia? The impact of communist norms and conventions, 1998–2004. *World Development* 38: 155–169.

Li, R. Y. M. 2006. Effectiveness of various construction safety measures in Hong Kong [BSc Thesis]. Hong Kong: The University of Hong Kong.

Li, R.Y.M., and S.W. Poon. 2009a. Workers' compensation for non-fatal construction accidents: Review of Hong Kong court cases. *Asian Social Science* 5: 15–24.

Li, R.Y.M., and S.W. Poon. 2009b. Future motivation in construction safety knowledge sharing by means of information technology in Hong Kong. *Journal of Applied Economic Sciences* 4: 457–472.

Li, R.Y.M. 2010a. Constructing sustainable vertical cities: Strategies to enhance closer cooperation between ASEAN contractors on pollution problem under the lens of economic game theories, cost benefit analysis and dynamic markov chain theories. *Asian Social Science* 6: 10–18.

Li, R. Y. M. 2010b. Knowledge sharing by web 2.0 in real estate and construction discipline. In E-research collaboration: Frameworks, tools and techniques, eds. M. Anandarajan and A. Anandarajan, 289–301. Berlin, Heidelberg: Springer-Verlag.

Li, R.Y.M. 2011. *Everyday life application of neo-institutional economics*. Saarbrucken: Lambert Academic Publishing.

Li, R.Y.M., and S.W. Poon. 2011. Using web 2.0 to share the knowledge of construction safety as a public good in nature among researchers: The fable of economic animals. *Economic Affair* 31: 73–79.

Li, R.Y.M. 2012. A content analysis in Hong Kong sustainable smart home knowledge sharing via world wide web. *Communications in Computer and Information Science* 350: 265–269.

Li, R.Y.M. 2013. The usage of automation system in smart home to provide a sustainable indoor environment: A content analysis in web 1.0. *International Journal of Smart Home* 7: 47–60.

Li, R.Y.M., and Y.L. Li. 2013. The role of competition law: An Asian perspective. *Asian Social Science* 9: 47–53.

Li, R. Y. M., and S. W. Poon. 2013a. Construction safety. Springer.

Li, R. Y. M., and S. W. Poon. 2013b. Demand for construction safety. In Construction safety, eds. R. Y. M. Li and S. W. Poon. Germany: Springer.

Li, R.Y.M., and S.W. Poon (eds.). 2013c. *A literature review on the causes of construction accidents*. Germany: Springer.

Li, R. Y. M., and S. W. Poon. 2013d. A literature review on the causes of construction accidents. In Construction safety, eds. R. Y. M. Li and S. W. Poon. Germany: Springer.

Li, R.Y.M., and S.W. Poon (eds.). 2013e. *Supply of safety measures in developing and developed countries: A global perspective*. Germany: Springer.

Li, R. Y. M. 2014. Law, economics and finance of the real estate market: A perspective of Hong Kong and Singapore. Berlin Springer.

Li, R. Y. M. 2015a. Construction safety knowledge sharing via smart phone apps and technologies. In Handbook of mobile teaching and learning, ed. Zhang Y, 1–11. Berlin, Heidelberg: Springer Berlin Heidelberg.

Li, R. Y. M. 2015b. Generation x and y's demand for homeownership in Hong Kong. Pacific Rim Real Estate Journal 21:15–36.

Li, R. Y. M. 2015c. Construction safety and waste management: An economic analysis. Germany: Springer.

Li, R.Y.M., S. Fong, and H.M. Chan. 2015a. Scholars' comments and criticisms on construction and municipal solid waste management in Hong Kong and Singapore from 2000 to present: A review of digital information online. *Construction safety and waste management: An economic analysis*, 31–45. Cham: Springer International Publishing.

Li, R.Y.M., S.W. Poon, and D.C.W. Ho. 2015b. A content analysis on static digital information about the causes of construction accidents. *Construction safety and waste management: An economic analysis*, 47–60. Cham: Springer International Publishing.

Li, R.Y.M., J. Zuo, and T. Fanen. 2015c. Supply and demand of construction safety regulations in developing countries. *Construction safety and waste management: An economic analysis*, 93–109. Cham: Springer International Publishing.

Li, R. Y. M. 2016. VR as an edutainment tool in secondary school sustainable development learning workshop. In Hong Kong Shue Yan University. Hong Kong.

Li, R.Y.M., and K.W. Chau. 2016. *Econometric analyses of international housing markets*. UK: Routledge.

Li, R.Y.M., H.C.Y. Li, C.K. Mak, and T.B. Tang. 2016. Sustainable smart home and home automation: Big data analytics approach. *International Journal of Smart Home* 10: 177–187.

Li, R. Y. M., and P. L. Ng. 2017. Wearable robotics and construction workers safety and health. In 8th International Conference on Applied Human Factors and Ergonomics Los Angeles.

Li, R.Y. M. 2017, forthcoming-a. 5d GIS virtual heritage. Procedia Computer Science 103.

Li, R. Y. M. 2017, forthcoming-b. Smart construction safety in road repairing works. Procedia Computer Science 103.

Li, R. Y. M., K. W. Chau, D. C. W Ho. 2017, forthcoming. Dynamic panel analysis on construction accidents in Hong Kong. Asian Journal of Law and Economics.

Lim, S., R.A. Buswell, T.T. Le, S.A. Austin, A.G.F. Gibb, and T. Thorpe. 2012. Developments in construction-scale additive manufacturing processes. *Automation in Construction* 21: 262–268.

Lin, J.-H.T. 2017. Fear in virtual reality (VR): Fear elements, coping reactions, immediate and next-day fright responses toward a survival horror zombie virtual reality game. *Computers in Human Behavior* 72: 350–361.

Lin, J., and A. Mills. 2001. Measuring the occupational health and safety performance of construction companies in Australia. *Facilities* 19: 131–139.

Ling, F.Y.Y., M. Liu, and Y.C. Woo. 2009. Construction fatalities in Singapore. *International Journal of Project Management* 27: 717–726.

Lingard, H., M. Hallowell, R. Salas, and P. Pirzadeh. 2017. Leading or lagging? Temporal analysis of safety indicators on a large infrastructure project. *Safety Science* 91: 206–220.

Liu, J., B. Li, B. Lin, and V. Nguyen. 2007. Key issues and challenges of risk management and insurance in China's construction industry: An empirical study. *Industrial Management & Data Systems* 107: 382–396.

Malekitabar, H., A. Ardeshir, M.H. Sebt, and R. Stouffs. 2016. Construction safety risk drivers: A BIM approach. *Safety Science* 82: 445–455.

Mitropoulos, P., T.S. Abdelhamid, and G.A. Howell. 2005. Systems model of construction accident causation. *Journal of Construction Engineering and Management* 131: 816–825.

Motter, A.A., and M. Santos. 2017. The importance of communication for the maintenance of health and safety in work operations in ports. *Safety Science* 96: 117–120.

Navon, R., and O. Kolton. 2006. Model for automated monitoring of fall hazards in building construction. *Journal of Construction Engineering & Management* 132: 733–740.

Park, C.S., and H.J. Kim. 2013. A framework for construction safety management and visualization system. *Automation in Construction* 33: 95–103.

Park, J., and H. Cai. 2017. Wbs-based dynamic multi-dimensional BIM database for total construction as-built documentation. *Automation in Construction* 77: 15–23.

Podgórski, D., K. Majchrzycka, A. Dąbrowska, G. Gralewicz, and M. Okrasa. 2017. Towards a conceptual framework of OSH risk management in smart working environments based on smart PPE, ambient intelligence and the internet of things technologies. *International Journal of Occupational Safety and Ergonomics* 23: 1–20.

Quest. 2017. Waiting for the internet of things? Iot is already here. Available: https://www.quest.com/ema-internet-of-things-infographic. Accessed 25 Apr 2017.

Rameezdeen, R., and A. Elmualim. 2017. The impact of heat waves on occurrence and severity of construction accidents. International Journal of Environmental Research and Public Health 14.

Raviv, G., B. Fishbain, and A. Shapira. 2017a. Analyzing risk factors in crane-related near-miss and accident reports. *Safety Science* 91: 192–205.

Raviv, G., A. Shapira, and F. Barak. 2017b. AHP-based analysis of the risk potential of safety indicators: Case study of cranes in the construction industry. *Safety Science* 91: 298–309.

Sacks, R., A. Perlman, and R. Barak. 2013. Construction safety training using immersive virtual reality. *Construction Management and Economics* 31: 1005–1017.

Sampaio, A.Z., and O.P. Martins. 2014. The application of virtual reality technology in the construction of bridge: The cantilever and incremental launching methods. *Automation in Construction* 37: 58–67.

Saurin, T.A., C.T. Formoso, and F.B. Cambraia. 2005. Analysis of a safety planning and control model from the human error perspective. *Engineering, Construction and Architectural Management* 12: 283–298.

Scott, J.T. 2009. Cost-benefit analysis for global public–private partnerships: An evaluation of the desirability of intergovernmental organizations entering into public–private partnerships. *The Journal of Technology Transfer* 34: 525–559.

Shehab, T. 2009. Automated measurement of earthwork volume in construction projects. *Construction Innovation* 9: 341–358.

Shin, M, H. S. Lee, M. Park, M. Moon, and S. Han. 2014a. A system dynamics approach for modeling construction workers' safety attitudes and behaviors. Accident Analysis and Prevention 68 95–105.

Shin, W, S. C. Jeong, and R. N. Lee. 2014b. A study on the causal analysis of death accidents by the falls in the construction sites. Journal of Korea Safety Management Science 16: 63–69.

Suárez-Cebador, M., J.C. Rubio-Romero, and A. López-Arquillos. 2014. Severity of electrical accidents in the construction industry in Spain. *Journal of Safety Research* 48: 63–70.

Teizer, J. 2016. Right-time vs real-time pro-active construction safety and health system architecture. *Construction Innovation* 16: 253–280.

Teo, E.A.L., F.Y.Y. Ling, and A.F.W. Chong. 2005. Framework for project managers to manage construction safety. *International Journal of Project Management* 23: 329–341.

Tinkercad. 2016. Tinkercad features. https://www.tinkercad.com/about/features.

Toole, T.M. 2002. Construction site safety roles. *Journal of Construction Engineering and Management* 128: 203–210.

Vysotskiy, A., S. Makarov, J. Zolotova, and E. Tuchkevich. 2015. Features of BIM implementation using autodesk software. *Procedia Engineering* 117: 1143–1152.

Wang, J., X. Wang, W. Shou, and B. Xu. 2014b. Integrating BIM and augmented reality for interactive architectural visualisation. *Construction Innovation* 14: 453–476.

Willmann, J., M. Knauss, T. Bonwetsch, A.A. Apolinarska, F. Gramazio, and M. Kohler. 2016. Robotic timber construction—expanding additive fabrication to new dimensions. *Automation in Construction* 61: 16–23.

Wu, M., A. Turpin, J. Thom, F. Scholer, and R. Wilkinson. 2013. Cost and benefit estimation of experts' mediation in an enterprise search. *Journal of the Association for Information Science and Technology* 65: 146–163.

Wu, P., J. Wang, and X. Wang. 2016. A critical review of the use of 3-d printing in the construction industry. *Automation in Construction* 68: 21–31.

Xia, M., and J. Sanjayan. 2016. Method of formulating geopolymer for 3d printing for construction applications. *Materials and Design* 110: 382–390.

Yu, Q.Z., L.Y. Ding, C. Zhou, and H.B. Luo. 2014. Analysis of factors influencing safety management for metro construction in China. *Accident Analysis and Prevention* 68: 131–138.

Yu, S.-N., B.-G. Ryu, S.-J. Lim, C.-J. Kim, M.-K. Kang, and C.-S. Han. 2009. Feasibility verification of brick-laying robot using manipulation trajectory and the laying pattern optimization. *Automation in Construction* 18: 644–655.

Zahoor, H., A.P.C. Chan, R. Gao, and W.P. Utama. 2017. The factors contributing to construction accidents in Pakistan: Their prioritization using the delphi technique. *Engineering, Construction and Architectural Management* 24: 1–39.

Zhang, S., J-K. Lee, M. Venugopal, J. Teizer, C. Eastman. 2013. Integrating BIM and safety: An automated rule-based checking system for safety planning and simulation.

Zhao, D., and J. Lucas. 2015. Virtual reality simulation for construction safety promotion. *International Journal of Injury Control and Safety Promotion* 22: 57–67.

Zhi, M., G.B. Hua, S.Q. Wang, and G. Ofori. 2003. Total factor productivity growth accounting in the construction industry of Singapore. *Construction Management and Economics* 21: 707–718.

Chapter 2
Robots for the Construction Industry

Abstract Employing robots on building sites may have been considered an unattainable fantasy in the past or simply a futuristic dream. Many construction practitioners are fiercely opposed to the use of robots and are wary of losing their jobs when such creations are intelligent enough to replace human resources. In contrast, the prohibitively high costs of manufacturing, producing and using robots on-site, in comparison to the costs of hiring labourers, is a major concern thwarting the ideas of many construction practitioners who wish to try something new and innovatory. Hence, hiring a robot to work on-site is not an imminent threat to the workers. Nevertheless, the recent technological breakthrough has caused waves in the industry and excited interest. It is reasonable to foresee that robots will usher in a new era in the construction industry. In this chapter, we adopt the data and method triangulation approach to study the construction practitioners', academics' and tool providers' viewpoints with regard to the costs and benefits of robots on construction safety and the construction industry. The interview results show that academics and construction practitioners in different parts of the world worry that robots may take jobs away from manual labourers. Wearable robotics have recently been introduced to one of the companies in Hong Kong's construction industry, yet most workers and even safety officers have no knowledge of this advance. A focus group interview has been conducted with a PowerPoint presentation and some research participants have worn the wearable robotics and commented on the tools' usefulness and efficacy.

Keywords Robot · Wearable robotics · Construction safety

1 Introduction: A General Overview on Robots

Between the 1960s and 1990s, many robots were designed for industrial application. They were used to rationalise manufacturing production and were equipped with prior task definition to execute works according to predefined programmes.

R.Y.M. Li, *An Economic Analysis on Automated Construction Safety*,
DOI 10.1007/978-981-10-5771-7_2

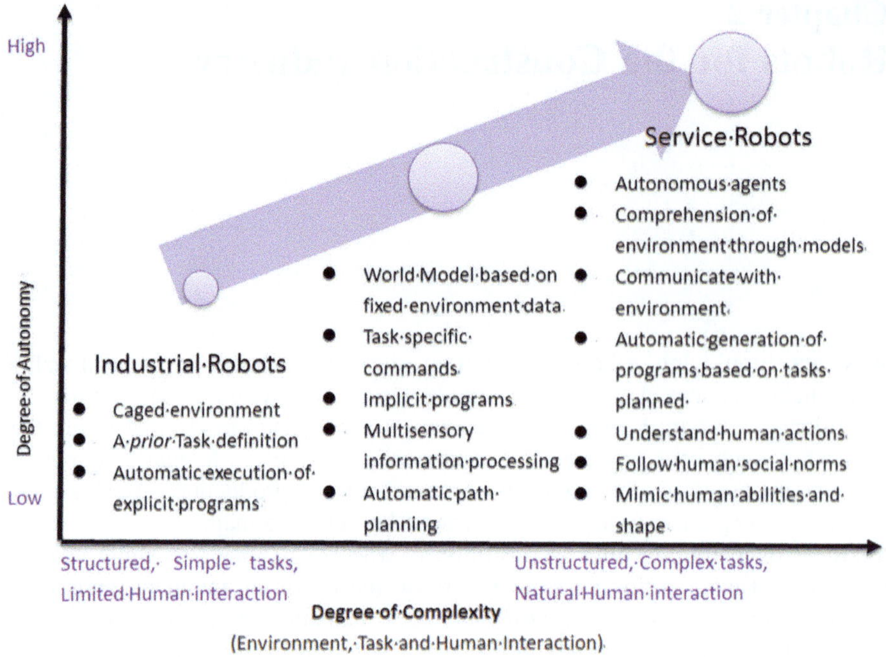

Fig. 1 Degree of autonomy and complexity from industrial robots to service robots (Haidegger et al. 2013)

Later on, robots were equipped with sensors to ascertain the working environment. Today, modern robots are ubiquitous. They have a certain level of artificial intelligence and can support, nurse and accompany humans (Haidegger et al. 2013; Li et al. 2016). Figure 1 illustrates the changes in robots over recent years.

In-depth research on robots suggests that they can autonomously react to any situation they encounter. For one example, they are able to open any type of door. Nevertheless, many of them can deal only with particular tasks in very specific situations (van Osch et al. 2014). Thus, it is often considered unrealistic to believe that a robot can fulfil some of the work tasks that have always been accomplished by human beings. Likewise, some construction practitioners do not wish to even consider the idea of adopting robots on-site for construction works at this premature stage.

No matter the perceptions, the reality or inconvenient truth is that robots are, in fact, slowly replacing some jobs originally conducted by humans. For example, some shopping malls use robots to replace ordinary security guards. A bank in China recruits a robot to act as receptionist "who" can handle customers' enquiries and persuade their potential customers to use their services. Recent research forecast that the US market will grow at 15% annually while the Chinese trade will increase by 17%. Service robots have become more significant throughout the first decade in the twenty-first century (Haidegger et al. 2013). As robots and other computer-assisted technologies take over manual labourers' tasks, there is rising concern about the future job market. By analysing the effect of the industrial robot usage from 1990 to 2007 in local labour markets in the United States, the research bolsters evidence that robots may reduce employment and wages. It is estimated that one more robot per thousand workers lowers the employment to population ratio by about 0.18–0.34% and wages by 0.25–0.5% (Acemoglu and Restrepo 2017).

2 Popularity in Robots, Robotic Arms and Wearable Robotic Searches as Reflected in Google Searches: A Big Data Analysis from 2004 to the Present

Whilst the subject of robots is discussed by many individuals, ranging from China to Peru, most of us may share the instinctive feelings that robots are envisaged as comprising two legs and arms only. In reality, there are three types of robots. The first is a traditional robot, with arms and legs, and able to move around freely. Smart home robots represent one of the very good examples. Another type is a robotic arm, which only processes arm-like movements (Fig. 2). Finally, there are now

Fig. 2 Robotic arm designed for industrial use, preparing food (photo taken by the author)

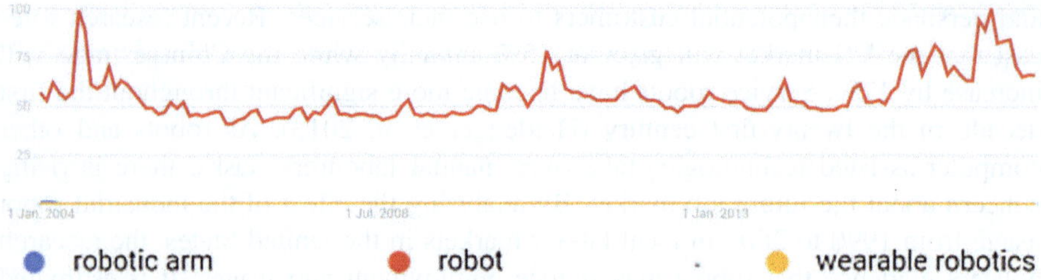

Fig. 3 Comparison with the number of searches between robot, robotic arm and wearable robotics from 1 Jan 2004 to 21 April 2017, in Google (2017)

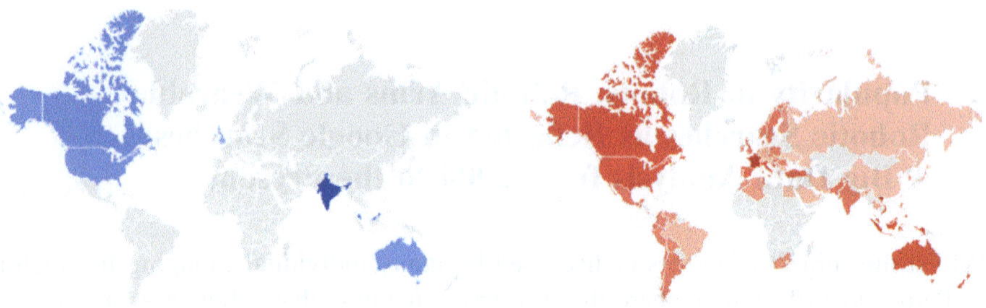

Fig. 4 Comparison of the regions with most searches on robotic arm (*left*) and robots (*right*) (Google 2017). *Note* There is insufficient data to illustrate regions for wearable robotics searches on the map (Google 2017)

wearable robotics, which need to be worn by humans to provide extra strength and support. All of these can potentially be used on sites (Li and Ng 2017).

Following the big data analytics adopted by Li et al. (2016), the figures below show the comparison the relative popularity of robotic arm, robot and wearable robotics in Google searches. There are a lot more searches in robot as compared to wearable robotics and robotic arm. The three largest number of searches of robotic arm comes from India, the Canada, the United States and Australia indicated by darker blue colour. On the other hand, the largest number of searches for robots are France, Bosnia and Herzegovina and Vietnam (Google 2017) (Figs. 3 and 4).

3 Information Flow Between Robot and Human

In general, there exists a core and domain ontology between developer and end user. The core ontology consists of sensors, actuators, drivers and operative systems and the end user domain ontology concerns the human–robot interaction. The

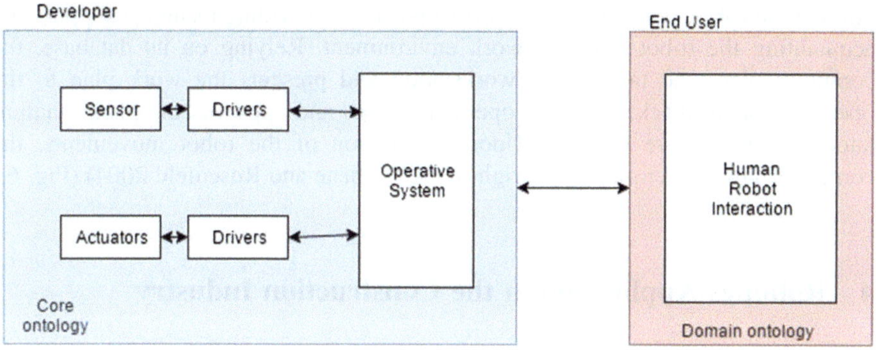

Fig. 5 Core and domain ontology between developer and end user (Haidegger et al. 2013)

human–robot system consists of three elements: computer, operator and robot. Figure 5: Core and domain ontology between developer and end user (Haidegger et al. 2013). A human operator performs two complementary actions: preparatory and supplementary work. The preparatory work includes data input to the computer with regards to the task. The supplementary work includes preparing the materials,

Fig. 6 The relationship between robots, computers and operators (Kahane and Rosenfeld 2004)

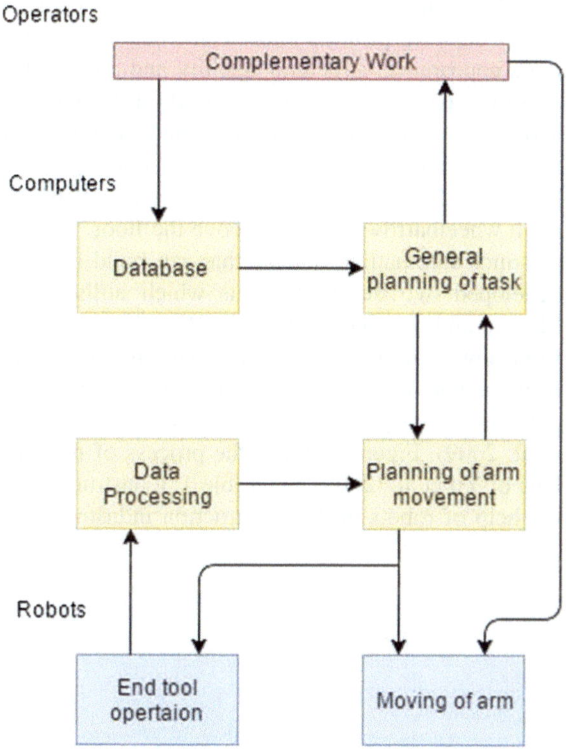

guiding the robot, placing it in the correct position, providing technical support and acquainting the robot with the work environment. Relying on its database, the computer plans the task at the workstation and presents the work plan to the operator, for feedback. After the operator revises and approves the plan, manipulator movements are initiated. Upon completion of the robot movements, the computer puts the sensors on the right track (Kahane and Rosenfeld 2004) (Fig. 6).

4 Robotics Application in the Construction Industry

Construction work usually takes place in a disorganised environment hosting many types and areas of danger. The idea of replacing construction workers with robots offers many advantages; for example, safety, quality and productivity improvement. Current trends in skyscraper towers, with accompanying escalations in labour costs and difficulty in hiring suitable workers, inevitably raises the need to adopt various robotic technologies in the construction industry (Jung et al. 2013).

The desire to move towards robot architecture has been developed through the current interest in constructing extremely high-rise buildings in Southeast Asia. Moreover, many of these structures do not take the form of traditional "straight up and down" designs but boast innovative typologies. Researchers have developed model-building robot systems that can construct 1:50 scale models. The model-building projects allow construction practitioners to envisage a new building's appearance, structural stability and construction reality (Bloss 2014).

At present, robots are mainly used for dangerous and laborious jobs. Bricklaying and paving is repetitive and hence can be conducted by robots. Despite its repetitive nature, bricklaying can lead to construction accidents. Bricklayers' workloads can be very high. The highest workload occurs when the bricks are loaded and unloaded via a wheelbarrow 0–50 cm above the floor. Anliker developed one of the earliest masonry automation systems that can build brick walls up to 8 m long. Lehtinen developed two masonry robots which utilise seam adhesive to glue the brick. Slocum and Schena produced a "Blockbot" that dry the stacks of concrete blocks. Pritschow et al. proposed a bricklaying robot that operates on-site to pick the bricks from prepared pallets, apply bonding materials and erect brickwork accurately, perform mortar plastering on the bricks and squeeze the brick onto the mortar (Yu et al. 2009). Figure 7 shows the process of making a bricklaying robot under the lens of Yu et al. (2009) and Table 1: Construction and infrastructure activities with the help of robots in the construction industry.

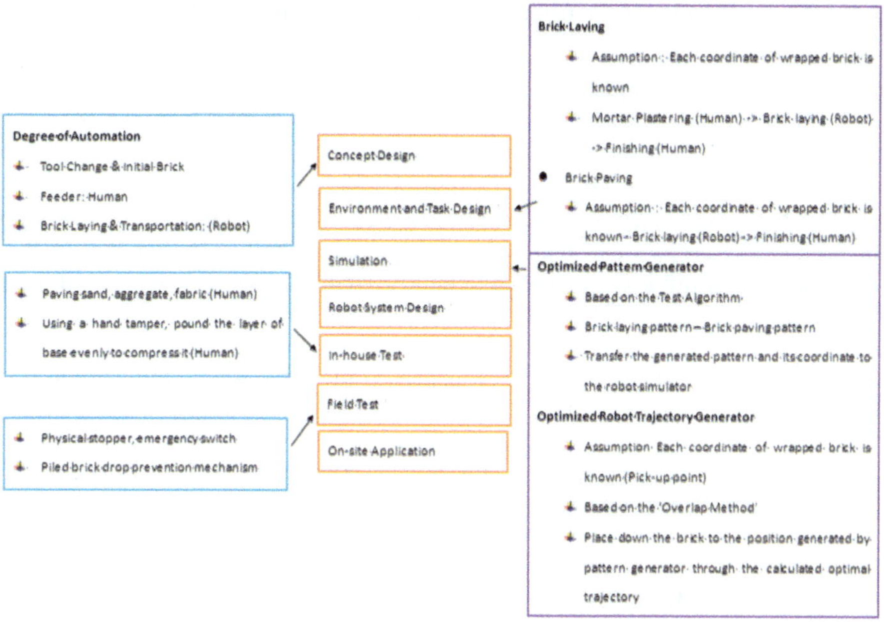

Fig. 7 The process of making a bricklaying robot (Yu et al. 2009)

Table 1 Construction and infrastructure activities with the help of robots (Bloss 2014)

Construction	Building services and maintenance
	Housing and building construction
	Construction in special areas, for example, deep sea, arctic zones, space and desert
Infrastructure	Bridge, container port, tunnelling and road construction
	Construction and deconstruction of power plants and dams

In view of these examples, gone are the days when robots could only be used in the design studio or laboratory. Many of the modern day robots address construction and fabrication issues on-site. These challenge the traditional construction approach and create new fabrication techniques for designers. In turn, this approach transforms architectural works by traditional construction techniques to feasible construction works on sites (Bloss 2014).

4.1 Four Types of Robots in the Construction Industry

4.1.1 Traditional Robots

Controlled by computers or other kinds of stimulus on-site, robots are used to autonomously construct the superstructure of buildings. Different types of robots are used for different types on sites. For instance, climbing robots have been used for bridge, skyscrapers and highways maintenance. The underwater construction robot for heavy work is driven by a hydraulic system that is robust to external force. Traffic Marshal Robot with motorised hand movement conveys car users a clear message that there is road work ahead. Likewise, the installation of heavy building materials, such as exterior curtain wall panels, is often hazardous and complicated: traditionally, a large amount of manpower is often needed (Li and Ng 2017). Indeed, with the help of robots, however, a number of workers who need to work on sites can be reduced. For example, PN Safety Industries (2017) has developed a Traffic Marshal Robot with motorised hand movement for road safety.

4.1.2 Wearable Robotics

The AWN-03 provides back support, senses a worker's motion and sends a signal to motors, which rotate the gears. The Suit AWN-03 embraces the user's shoulder, waist and thigh. In essence, it assists construction workers' movement when they lift and grasp heavy items. It raises workers' upper body support, pushes their thighs and lessens lower back stress by 15 kg. The battery power pack of AWN-03 lasts for six hours and each of the robot suits is sold for approximately US$8100. It is expected that there will be an increase in demand for AWN-03 amid the labour shortage problems and the ageing workers in the construction industries (Li and Ng 2017).

FORTIS, another wearable exoskeleton, enhances users' strength and work clothing. Similar to Iron Man, the tool is unpowered and light in weight. The external structure enhances the user's endurance. It aids workers in lifting heavy loads, such as reinforcing bars, and while using industrial tools. It transfers loads to the ground via the exoskeleton when the construction workers stand or kneel. It creates a weightlessness sensation when wearers are carrying or manoeuvering heavy objects. The exoskeleton's ergonomic design moves naturally with the wearer and is able to adapt to various different body heights and types. Capable of supporting up to a thirty-six-pound instrument, it is designed like the bucket of a cherry picker or a man lift. It is used to support large tools which may be tiring to operate overhead and horizontally, such as grinders, demo hammers, rivet busters, etc. (Li and Ng 2017).

4.1.3 Robotic Arm

Robotic arms are usually discussed together with robots as their size is relatively small but they can carry out all types of work conducted by robots. For example, a robotic arm is constructed with servo brackets, which are made of aluminium due to its lightweight properties but are stiff, to mimic a human arm. Similarly, the robot gripper is also made of aluminium for the same reason (Candelas et al. 2015).

Figure 8 shows the design of a typical robotic arm. A robotic arm includes an infrared sensor to determine radial distance, a USB camera to detect the necessary accelerators, angular orientation to provide feedback for angles and force sensors to determine whether the arm can grab an object. The constraints of the robotic arm can be solved by Optimisation Toolbox in Matlab via fmincon and other software. Table 2 demonstrates the sensors required to operate the robotic arm, and Fig. 9 shows the typical mini-robotic arm with gripper, servos (to alter the direction of the robotic arm) and Arduino board (acts as the "brain" of the robotic arm) (Li 2017, forthcoming-b).

Fig. 8 Design of robotic arm (Li 2017, forthcoming-b)

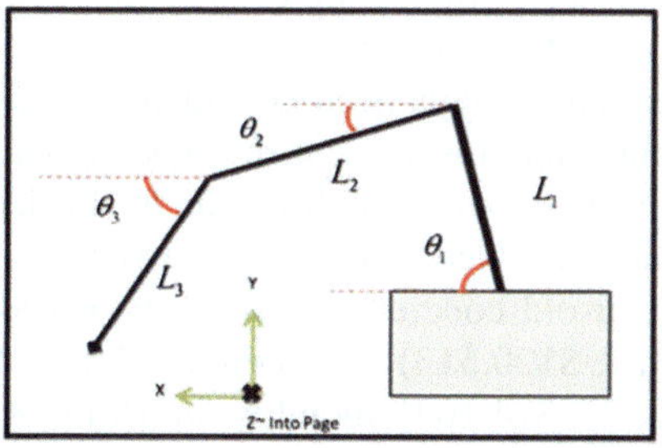

$$X = \sum_{i=1}^{3} \cos(\theta_i) L_i$$

$$Y = \sum_{i=1}^{3} \sin(\theta_i) L_i$$

Table 2 Sensors required for making the robotic arm (Li 2017, forthcoming-b)

Sensors needed	Reasons
Two infrared sensors	To determine the radial distance between the object and the arm
USB camera	To detect the angular orientation of the object
Accelerators	To provide feedback for angles $\theta 1, \theta 2, \theta 3$
Force sensor	To determine if the arm grasps something

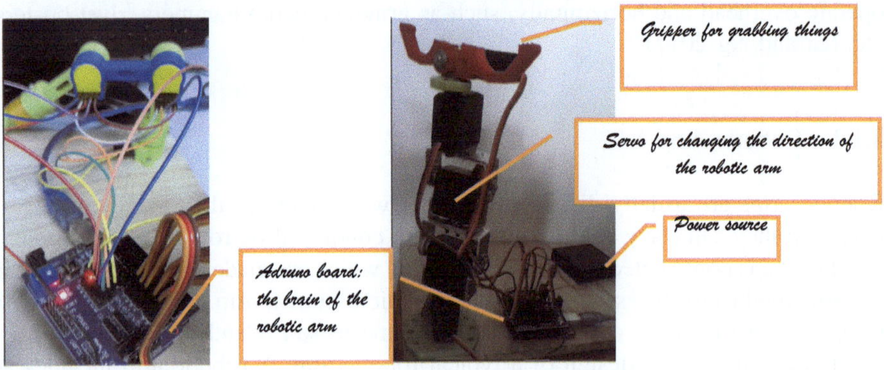

Fig. 9 A robotic arm made by the author (author's photo) (Li 2017, forthcoming-b)

5 Monetary Benefits of Using Robots on Sites

"A penny saved is a penny earned." When making decisions to purchase materials that are new to the market, most companies need to recognise the benefits they will incur when they make any groundbreaking decision. Robot purchase is no exception. Senior management personnel need to be certain that such purchases make economic sense for the future of the company.

Revisiting the issue of the nature of the construction site and the inherent labour shortage, a clear picture emerges regarding why robots are adopted. The construction industry has always been considered as dangerous. The heat of summer and the cold winter months worsen the work environment in the cities and deter many workers from seeking employment in the industry. In view of the labour shortages, manual labourers' costs are inexorably rising around the globe. For example, a 27-year-old concrete worker toils at least 18 h per day and earns HK $160,000 (about US$20,513) per month. Many of the workers even earn HK $110,000 (about US$14,103) monthly in Hong Kong (see Table 3). The high labour costs, coupled with possible compensation payouts, have become two of the major reasons why construction industry practitioners are exploring different outcomes through employing robots on sites.

Table 3 Construction workers' salaries in different trades in Hong Kong (Yu 2016)

Types of work	2015/2016 (HK$ per day) (HK$7.8 = US$1)	2016/2017 (HK$ per day) (HK$7.8 = US$1)
Plasterer	$1350	$1450
Carpenter	$1130	$1230
Painter	$1120	$1170
Plumber	$1250	$1350
Formworkers	$2050	$2500
Steel fixer	$1930	$2150
Concreter	$2300	$2500
Scaffolder	$1700	$1800
Surveyor	$1200	$1300
Electrician	$1150	$1210
Metal Scaffolder	$1150	$1250
Metal worker	$1080	$1150
Mason worker	$1100	$1150
carpenter	$1130	$1230
Ceiling and Partition labourer	$880	$1000
Pile driver	$1300	$1350

6 Are Robots Safe to Use? Safety Issues in Using Robots

Asimov mentioned robot safety as early as 1942. However, Asimov's three laws are insufficient to cope with today's robots' behaviour. Modern industrial robot standards provide guidelines for robots and their working environment. ISO 10218-1 (2011) and ISO 10218-2 (2011) standards delineate the requirements to lessen critical hazards. Project teams need to consider the trade-off between technical attributes and pertinent safety requirements, due to the emerging issue of the robot. Interlocks and guards protect people who work with robots. For years, industrial robot safety has been considered as a problem, in use, and has resulted in laws and regulations which keep users away from robots. Well-defined operations, such as programming, repair and teaching, are difficult (Mitka et al. 2012). The American National Institute of Standards and Technology classifies three levels of safety zones for robots (Mitka et al. 2012) as shown in Fig. 10. Figure 11 demonstrates the risk assessments and reductions for robots on construction sites (Mitka et al. 2012).

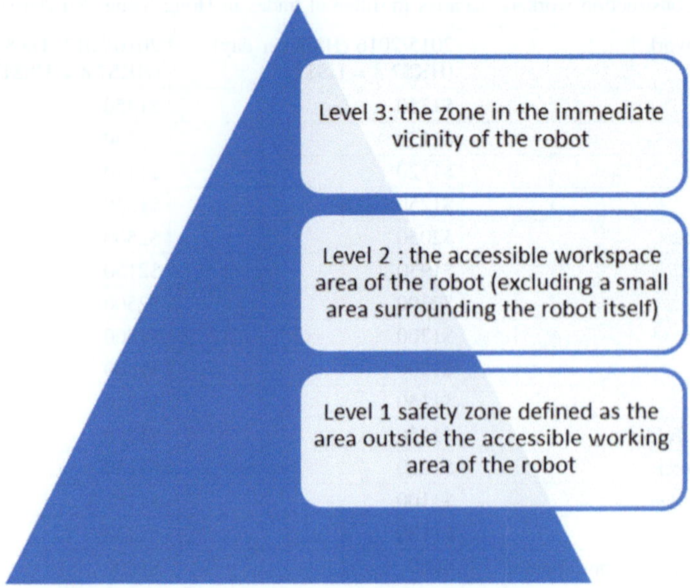

Fig. 10 Three levels of safety zones for robots (Mitka et al. 2012)

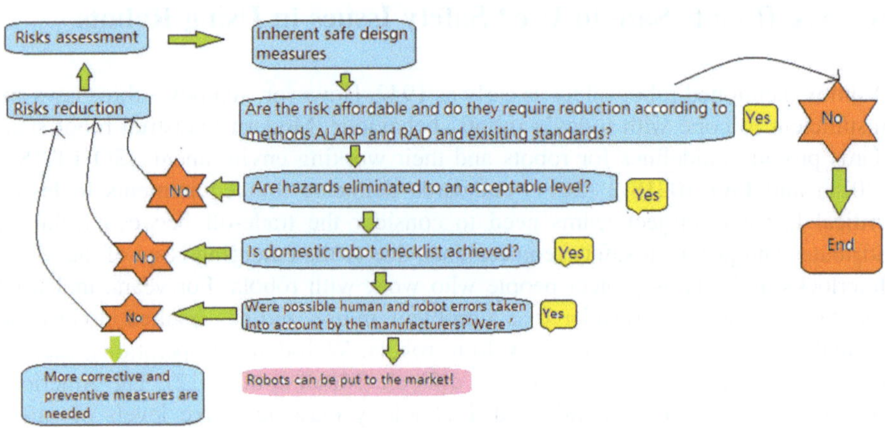

Fig. 11 Risk assessments and reductions for robots on sites (Mitka et al. 2012)

7 Research Method

In this chapter, we adopt data and method triangulation to study various stakeholders' viewpoints on robots, robotic arm and wearable robotics' implication on construction safety. Previous research results suggest that different research methods have their weaknesses and strengths. Method triangulation research (a combination of different research methods, between two or more) enhances accuracy and validity (Li 2015).

We first conduct an in-depth interview with construction practitioners. Following that, we conduct a case study. As some of the robotics, such as robotic arms and wearable robotics, had not yet been introduced to Hong Kong by the time we conducted the case study, focus group interviews where the interviewees can try the tools and presentations are particularly useful in studying the practitioners' viewpoints and perceptions.

8 Practitioners Viewpoints on Robots

In Hong Kong, many of the interviewees are aware that the usage of robots on sites is imminent. An innovation manager in a large-scale construction company mentions that his company has used robots for installing window panels. One of the CEOs of a construction firm suggests that their company has adopted the robots for laying pipes on construction sites, as it is difficult for workers to do so underground. A safety officer in a gas company suggests that when the existing gas conduit for the construction sites is repaired, his company does not want to dig the ground and repair it. Robots can be used for repairing the existing gas conduit without causing much disruption to the nearby neighbours. Thus, we may conclude that in Hong Kong, most of the robots are used to perform varying jobs that prove difficult for manual workers to undertake, such as underground work and heavy-duty works, including window panel installation.

The benefits of robots may be addressed by the question posed by a CEO in one of the largest construction companies in Seattle: "[*the question that we have in the construction industry is*] *how do we leverage people? We have so many repetitive* [*tasks*] *that need to be done manually…*" A section chief for a construction company in Fujian, China, suggests that robot manipulators enable large-scale factory automation of simple and repeated tasks. A project manager, currently stationed in Shanghai, agrees that robots can save time and offer high efficiency.

8.1 Implications of Robots on Construction Safety

Most of the interviewees agree that the reason why the employment of robots must lead to fewer accidents is prominently due to their ability to replace manual workers. For example, a structural engineer in Hong Kong suggests that "*it is in no doubt that a robot is helpful in construction. Robots play an important role in taking down a wall, which can replace construction workers and reduce the injury rate. Also, it can forecast the difficulties when constructing a building*". A labour union chairman in Hong Kong suggests that "*robots can replace workers in collecting and putting the safety cones in the road. As many of the recent construction workers were killed or injured due to reckless drivers carelessly hitting the workers, replacing the manual work by robots can reduce safety risks. Likewise, the robots can also work underground where the confined space is so small that there is no room for humans to work and unknown toxic gases may cause accidents*". A CEO of a safety investigation company suggests that "*robots can work at hazardous jobs; for example, they can work in environments with asbestos*". He also quoted an example of when a fire exploded in a warehouse in Hong Kong, causing the death of two firemen. A client of his asked if their company could send their robots to take photos. Thus, he considers that robots can act as helpful assistants in planning ahead for some of the dangerous jobs. Nevertheless, he also agrees that not all work can be replaced by robots, despite the safety perspective. For example, we need to hire a worker to monitor the robots. It is hard to hire a robot to monitor the robots. With given and limited resources, humans have to make their own judgements on whether robots can be used for the sake of safety and other concerns.

An academic researcher (also working as an architect) from Hong Kong shares similar thoughts and suggests that the ability of robots to reduce construction accidents is mainly due to the fact that "*the dangerous work or work with high risk could be carried out by* [*them*]". Two professors from different universities in the United States are also in agreement. One of them suggests that "*the costs will be high but the benefits could save lives*". Another professor proposes that "*as there are fewer humans on sites, there will be lower chance for human injury*".

8.2 Costs of Using Robots

Most of the construction practitioners are sceptical about adopting robots on sites. Some of them consider that the major drawbacks for robots are the financial considerations. They are of the view that the costs for adopting robots on sites are higher than other automated tools, such as virtual reality and additive printing. An academic professor from Melbourne reveals that the "*costs include robots, software and hardware engineers for customised applications for the robots*". A university professor in the United States who still holds an industry position in Iran concedes that a "*robot is expensive technology and needs training and expertise*". A CEO of a construction company in Guangdong added "*the cost is the price and technology. The price of making the robot in a safe way is very expensive. Also, it needs in-depth technology to maintain the safety*".

A senior management from a construction company in Seattle, however, does not agree with them. He holds an optimistic view on this issue as his company is now investing a large amount of resources in developing artificial intelligence, meaning that the robots can have their own “thoughts”, making the best decisions for the building processes, saving time and money for the construction company. Thus, they consider what they are doing now is a good investment. A robot provider suggests that “*despite the costs of buying robots, it saves the labour costs a lot, with higher efficiency since it does not need a break; just needs electricity then it works 24/7*”. A senior management member of a safety investigation company in Hong Kong is of the view that the costs are only high at the initial stage but will reduce as the tool becomes mature and more widely used.

A top construction journal editor in chief in the United States also throws light on artificial intelligence of robots “*robots are mostly used for surface furnishing, simple for robots to do, but dangerous for humans to do. Robots must have artificial intelligence to accomplish the tasks. Nevertheless, if that has too much AI, they are very expensive. Most robots have to be as cheap as possible with less AI and accomplish the work*”.

8.3 Robots Replacing Manual Workers on Sites Is Simply a Fantasy

A CEO of a construction firm in Hong Kong keeps things in perspective and suggests that “*robots can hardly replace human work, because the robots cannot climb as a human climbs. On the other hand, the robots are so heavy that the scaffolding cannot support [the heavy robots] and robots cannot work well with some of the small tools*”. Another safety officer who works for one of the largest contractors concedes that “*the robots cannot replace human beings, as artificial intelligence is not that well-developed*”.

> I think robots are difficult to work on site. Factory-like production can be done by robots as works are fixed and can be easily standardised. Nevertheless, it is quite difficult to use them on sites when works are not fixed. Every single piece of work can have errors: we have to make sure everything is in a correct place. You cannot make a minor mistake as you cannot pass the standard if there is any error.
>
> Because a construction site is not flat enough, robots may not work well. When it comes to building flats, it will not even be flat. This is because there will be more job mud. When we build flats, there will be more materials needed. When there are over 20 different types of material on sites, it will be more difficult.

A safety officer in Macau suggests “*I think this can have a use in planning or method statements, such as for display. For example, we can use robots to show the installation of buildings, which can be used in the long-term or existing situations. I think using a robot is the most useful*”.

A structural engineer thinks that robots cannot replace all the manual labourers' work: "*There is still a question if the robot is an end solution, since its price is too high and it is in doubt whether today's technology can deal with a variety of jobs on sites... Whether robots [are useful or not for construction works] depends on the [nature of the] job. Some manual work is impossible to be replaced by robots. For example, it is hard to find a robot to carry the tool and weld for you*".

A professor from the United States with over thirty years of research in robotics comments "*the number one reason for not using robots is construction law preventing [robot] from happen. Nobody wants to be number one in the industry. Everybody wants to be number two. They know all the risks. Risks mean we have product liability, you have insurance policies for contract design...if you going to use this technology, we (insurance) are not going to cover it. It is unprovened, it is untested and therefore you are entering to construction liability and product liability. And there is professional liability...Legal liability besides contract law that the unprovened technology cannot be used in a given site. You have professional liability that extends to people who make decision, either from design to those who take deicision*".

8.4 Are Robots Threats to Construction Workers?

Adopting robots for various kinds of jobs should come as news to no one. One of the traditional explanations for why we do not adopt robots for work is due to everyone's natural fear of losing jobs. Thus, wider adoption of robots on sites is still a moot point. Some interviewees consider it is impractical and unworkable to use robots on sites and that the idea has no merit. A researcher working in Switzerland is worried that robots will worsen the current high unemployment rate in Europe: "*If I am an employer, I cut costs to [reduce] intensive labour: it is ok in one enterprise or so. If we all do that, it can be problematic. If we put drones in as well, they do the most dangerous work. People do not need to do the risky work. The big picture...many people will be unemployed if digitalisation can be shared by the population as a whole*".

> In Spain, 25% are unemployed. Actually more than this. They have difficulty finding a job in the market. Some employed subsidise those who are unemployed. Limited extent of migrants and very controlled. If you find the way to distribute a mining site, digging out … generate some wealth at least to a minimum, to lose the job, it's better to have robots to work for digging something in the sea. The question is how to deal with the issue. That's better. Look at Arab Spring who don't see the chance in the labour market riots and social unrest.

Thus, he suggested that the major hurdle in hiring robots lies mainly in Trade Unions. "*The trade union is very powerful and manages to put some limits on technological change. They have strong working hours and we work more on developing countries. You don't need so much support. If in a market approach, private companies will try to cut costs as much as possible. I don't know whether we should stop it. The private companies are investing but Switzerland is a small country. The economy is not growing fast. They benefit from international. Everybody will be employed. The question is what happens if the welfare is not there. That's a question we need to answer together. What happens if Europe subsidises those who do not work. The number of unemployed is growing.*

9 Modern Application of Robots, Robotic Arm and Wearable Robots on Sites

9.1 Application of Robots on Sites

One of the largest Hong Kong local construction firms has adopted robots to reduce the number of workers and to enhance productivity and safety. In one of the hotel redevelopment projects, the presence of robots reduced the number of workers on-site by 25% (Figs. 12 and 13).

Fig. 12 Large window panels are installed with the help of robots (author's photo)

Fig. 13 Large window panels are installed with the help of robots (author's photo)

9.2 Application of Wearable Robotics

This construction company has also purchased two sets of exoskeletons, a kind of wearable robotics from Japan that protects their workers, improve their efficiency and reduce risk. According to the company's safety officer, it is used to protect workers' backs and waists. This is particularly useful when they work for a prolonged period. However, the cost of this technology is high: it is difficult for a small-size company to use such innovations (Li and Ng 2017).

As wearable robotics, such as exoskeletons and robotic arms, reduce strain and lighten heavy loads, they inevitably benefit older workers. As robotics technology continues to develop, coupled with the rapid progress in programming and robot maintenance techniques, this certainly opens up new opportunities for younger staff, who are often considered to be the best age group to welcome modern technology the most and who often look for innovative applications, such as wearable robots. Nevertheless, some workers worry that robots will eventually take jobs away from them or lead to some of the scenarios from a science fiction film. Yet, many countries and cities in the developed world experience the problem of labour shortage as fewer people enter the industry. Therefore, robots are simply doing their jobs in filling the gap left by a dwindling workforce (Li and Ng 2017).

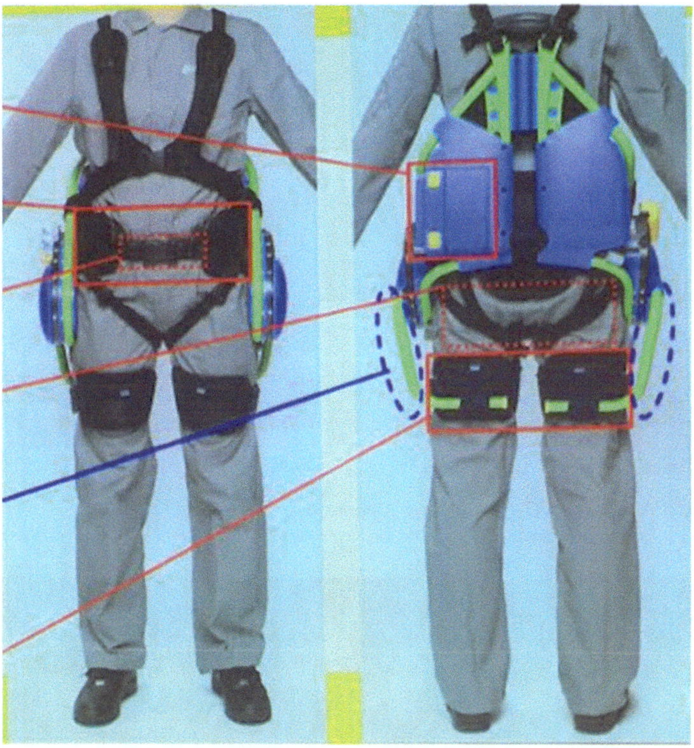

Fig. 14 Panasonic ActiveLink AWN-03 ("Ekso works," 2016) (author's photo)

In Hong Kong, using the robot's motor assists with human body mechanics, minimising the risk of human error. It reduces physical strain during human work, assists lower back body movements, detects the user's motion in the lower back when lifting and holding heavy objects and then sends a signal to the motors to rotate the machine's gear. It raises workers' upper body whilst pushing their thighs, lowering the user's stress at the back by 15 kg. Users can wear the tool by themselves and do not require extra help from others. Thus, it is easy to use. It has three major features: an auto assist mechanism, it is easy to put on and the battery can last for 8 h. https://www.youtube.com/watch?v=zGmymin7d0o (Figs. 14, 15 and 16).

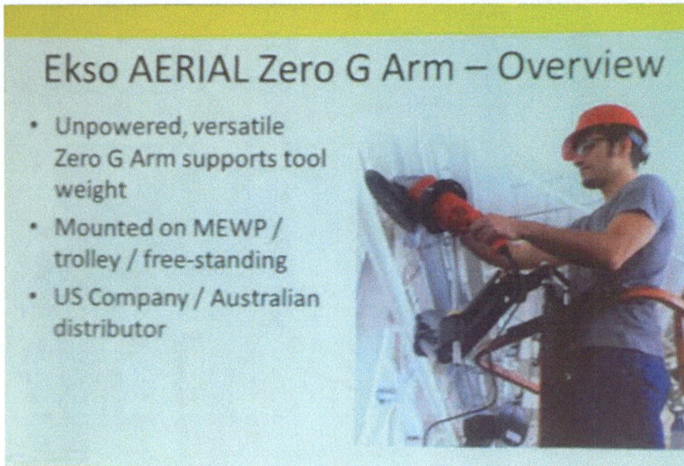

Fig. 15 Ekso Aerial Zero G Arm (author's photo)

Fig. 16 Ekso Aerial Zero G Arm (author's photo)

9.3 *Focus Group Interview Results of Practitioners' Perspectives on the Exoskeleton*

In this research, we were grateful to have the opportunity to attend the lunch meeting offered by a construction company to conduct a focus group interview. The meeting began with a detailed PowerPoint introduction on wearable robotics; i.e. various types of exoskeleton presentation. After that, workers and our research assistants tried to use the tools.

Worker A suggests that the company may purchase some exoskeletons: when the workers need the tool for one particular procedure, they can get one from stock. The costs are quite high currently but will soon decrease when they gain popularity. Worker B recalls that at the very beginning, the costs of mobile phones were quite high. The costs of the exoskeleton may also reduce substantially, later, but he thinks that will take a very long time. At the meeting two workers were also invited to wear the exoskeleton.

Worker C, a well-built man, suggests that the tools should be able to help him in moving heavy equipment. Nevertheless, another worker states that the person who wears the exoskeleton can only help him in moving heavy materials if his hand is

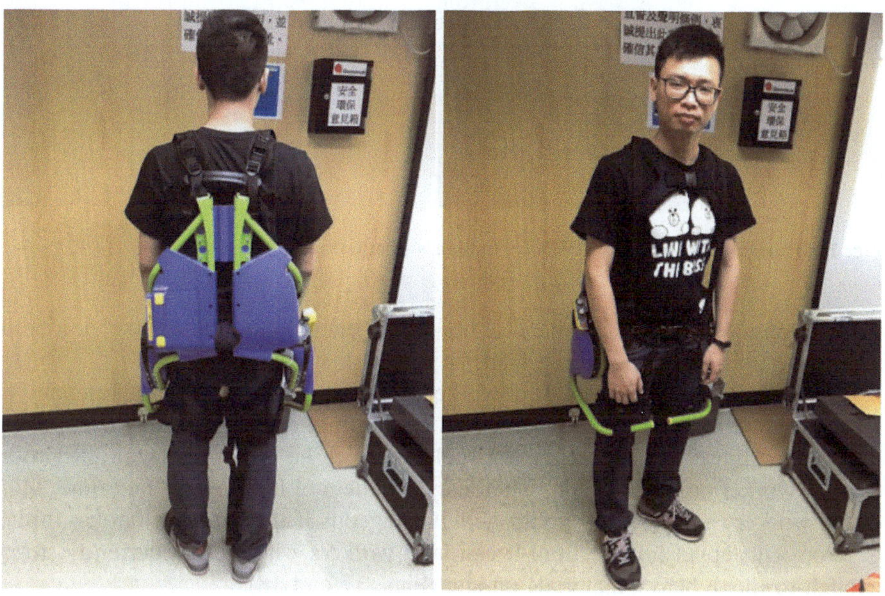

Fig. 17 AWN-03 robotic (author's photo)

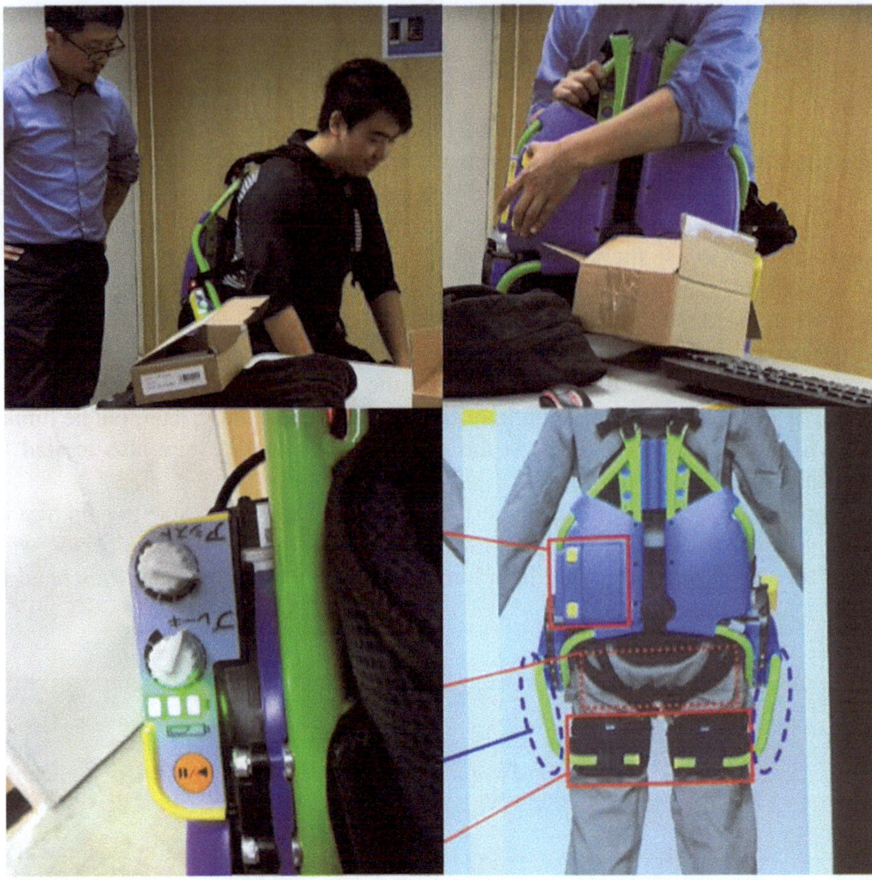

Fig. 18 Panasonic ActiveLink AWN-03 (author's photo)

really strong. If the workers' hand is not strong enough and the exoskeleton helps him move up his body with the load, he may damage his hands.

Thus, we invited a slightly built worker (D) to try to use the exoskeleton and he also conceded that the exoskeleton could help him in carrying heavy blocks. Even the worker himself can wear the exoskeleton, once he is familiar with the procedure. A worker only needs help when they use the tool for the very first time. Most of the workers agree that it would be advantageous if the exoskeleton can further develop with upper arm section. It can help workers who are not currently strong enough to move heavy materials on-site (Figs. 17 and 18).

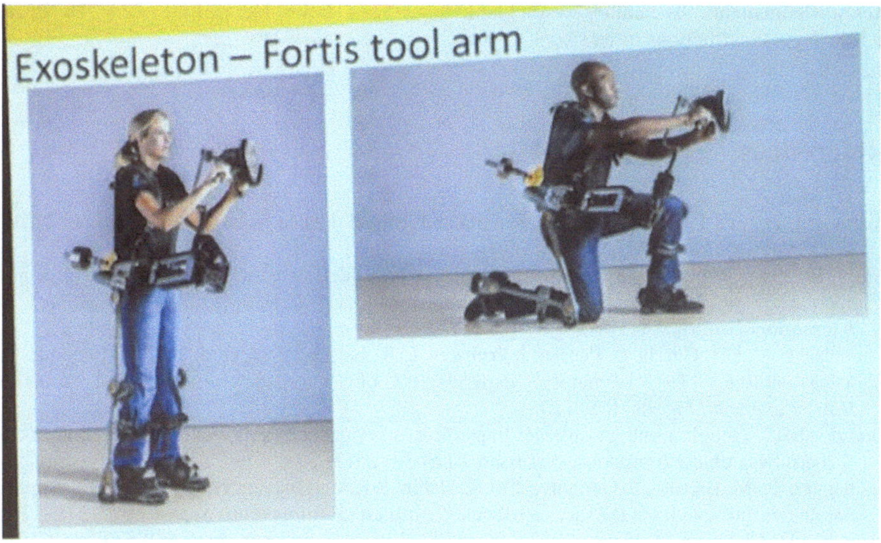

Fig. 19 Exoskeleton Fortis tool arm (author's photo)

With regard to the Fortis tool arm, Worker A considers that the arm can help to do such work as drilling, when the activity must be sustained for a long period. This can provide much extra arm strength and prevent fatigue (Fig. 19).

10 Conclusion

Among the many companies that engaged in the interviews, only the largest scale companies have adopted robots for works on sites. Usage of robots sounds as though they represent status symbols for wealthy, large-scale construction companies. Full automation by robots on sites is still work-in-progress. This may be due to the relative high costs at present. Whilst the costs of making robots are not that expensive (sensors, motors, electric wires are relatively cheap), there are quite high costs incurred in doing research and development before robots can be usable on sites. Nevertheless, it is reasonable to foresee that the number of construction companies adopting robots on sites will increase. This is particularly true when the costs of hiring manual labour increase but the costs of robots decrease due to mass production. Thus, is it worth taking the decision to invest in such an innovative approach? Whilst some of the stakeholders may consider this whole development as a new technology currently in its very early stages, the outcome should be positive in the near future.

Acknowledgements The author would like to thank Miss H. Siu, Mr. H. Li, F. Zeng, D. Ng and D.K.F. Ng's contributions to this book chapter.

References

Acemoglu, D., and P. Restrepo. 2017. Robots and jobs: Evidence from us labor markets. NBER Working Paper No 23285:1–91.

Bloss, R. 2014. Robots have come to architecture to model, construct, fabricate and offer new approaches to create innovative designs, elements and structures. *Industrial Robot: An International Journal* 41: 403–407.

Candelas, F.A., G.J. García, S. Puente, J. Pomares, C.A. Jara, J. Pérez, et al. 2015. Experiences on using arduino for laboratory experiments of automatic control and robotics. *IFAC-PapersOnLine* 48: 105–110.

Google. 2017. Google trend. Available: https://trends.google.com/trends/explore?q=robot,robotic%20arm,wearable%20robotics. Accessed April 30, 2017.

Haidegger, T., M. Barreto, P. Gonçalves, M.K. Habib, S.K.V. Ragavan, H. Li, et al. 2013. Applied ontologies and standards for service robots. *Robotics and Autonomous Systems* 61: 1215–1223.

Jung, K., B. Chu, and D. Hong. 2013. Robot-based construction automation: An application to steel beam assembly (part ii). *Automation in Construction* 32: 62–79.

Kahane, B., and Y. Rosenfeld. 2004. Real-time "sense-and-act" operation for construction robots. *Automation in Construction* 13: 751–764.

Li, R.Y.M. 2015. Generation x and y's demand for homeownership in hong kong. *Pacific Rim Real Estate Journal* 21: 15–36.

Li, R.Y.M., H.C.Y. Li, C.K. Mak, T. Tang, and Beiqi. 2016. Sustainable smart home and home automation: Big data analytics approach. *International Journal of Smart Home* 10: 177–187.

Li, R.Y.M., and P.L. Ng. 2017. Wearable robotics and construction workers safety and health. In: 8th International Conference on Applied Human Factors and Ergonomics Los Angeles.

Li, R.Y.M. 2017. Forthcoming. Smart construction safety in road repairing works. *Procedia Computer Science* 103.

Mitka, E., A. Gasteratos, N. Kyriakoulis, and S.G. Mouroutsos. 2012. Safety certification requirements for domestic robots. *Safety Science* 50: 1888–1897.

PN Safety Indutries. 2017. Pn safety indutries. Available: http://www.pnsafetyind.com/. Accessed April 2, 2017.

van Osch, M., D. Bera, K. van Hee, Y. Koks, and H. Zeegers. 2014. Tele-operated service robots: Rose. *Automation in Construction* 39: 152–160.

Yu, C.T. 2016. The construction institute association mentioned that the abovementioned items are daily wage of the workers. Available: http://www.hk01.com/sns/article/48907?utm_content=buffer99389&utm_medium=Social&utm_source=facebook%20hk01&utm_campaign=buffer. Accessed October 18, 2016.

Yu, S.-N., B.-G. Ryu, S.-J. Lim, C.-J. Kim, M.-K. Kang, and C.-S. Han. 2009. Feasibility verification of brick-laying robot using manipulation trajectory and the laying pattern optimization. *Automation in Construction* 18: 644–655.

Chapter 3
Building Information Modelling and Construction Safety

Abstract Many construction practitioners may have felt the pinch of information technology in recent decades. Building Information Modelling is one of the most useful pieces of software for people in this profession, as it helps them to foresee a building structure before it is constructed. During construction, BIM can let us see clearly the procedure and process of different parts of creating a structure in different times; hence, it saves the costs of making a building model in physical form. On the other hand, it helps retrofit companies to foresee the facility to be placed in a room before we put all the furniture inside. That essentially enhances the communication between different stakeholders. Besides this, usage of BIM improves safety. The results of our case study suggest that the software fit like a glove with additive manufacturing. This paper casts light on the fact that more information technology and construction practitioners may recognise the importance of BIM in safety at the end of the day. In view of all the above-mentioned advantages brought by BIM, it is high time for construction practitioners to pin their ears back to fully utilise it for safety and cost saving.

Keywords BIM · Construction safety · Costs and benefits analysis

1 Introduction

Gone are the days when the lack of skills tied construction practitioners hand and foot in applying information technology on site. Technological breakthroughs in recent years have changed the façade of the construction industry root and branch, and paper and pens are no longer the only means of showing the architectural designs of building and civil engineering structures. The history of BIM can be dated back to the late 1950s, when the United States defence contractor Itek Corporation introduced a computer graphics system for engineering designs. This subsequently changed to the three-dimensional BIM, which allows individuals to visualise the X–Y–Z dimensions of a structure (Ghaffarianhoseini et al. 2017).

R.Y.M. Li, *An Economic Analysis on Automated Construction Safety*,
DOI 10.1007/978-981-10-5771-7_3

A modern BIM model allows us to construct the building structure virtually, thereby detecting clashes and ensuring an optimal construction sequence in a way that is not feasible in using paper-based representations. Where BIM truly shines, however, is in its recent form, which allows for augmentation of the three-dimensional space with time, facility management and cost (Howard et al. 2017),

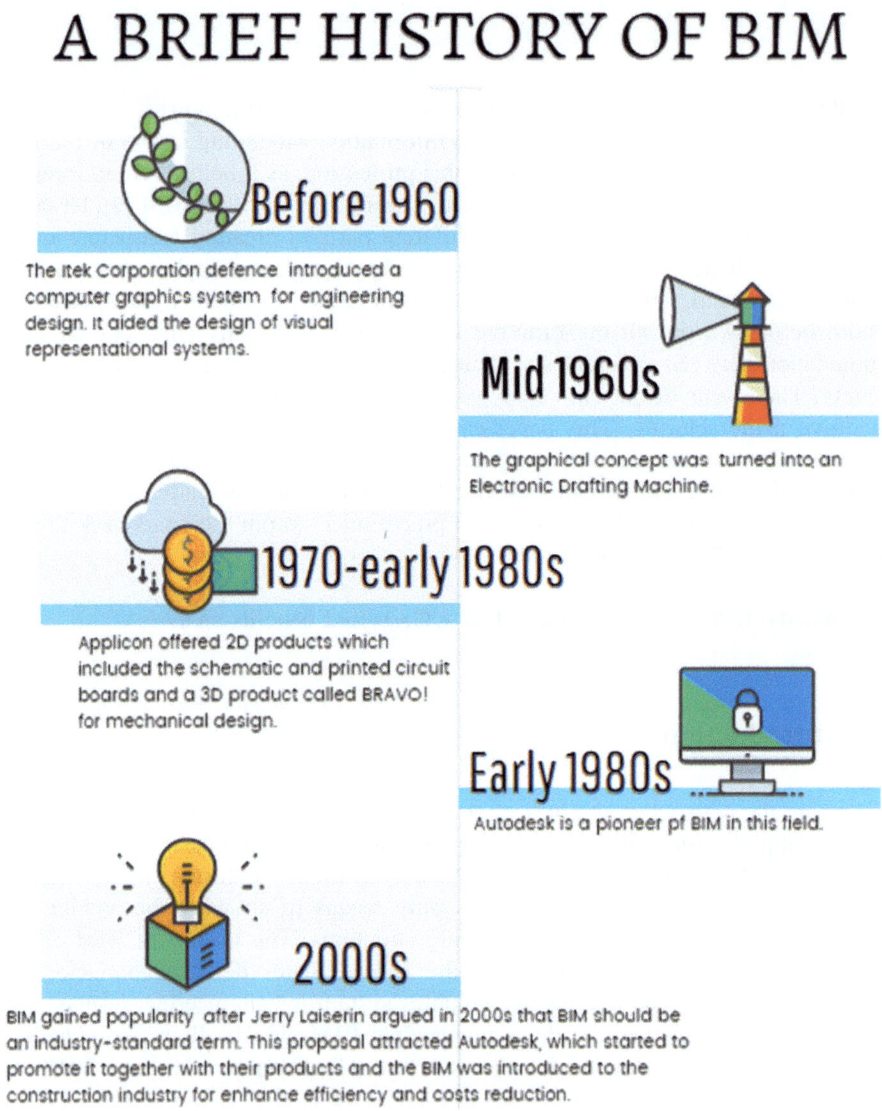

Fig. 1 A brief history of BIM (Ghaffarianhoseini et al. 2017) (author's figure)

construction scheduling and progress tracking, structural analysis, building energy analysis and job safety in AEC computerised platforms (Ghaffarianhoseini et al. 2017). Figure 1 illustrates a brief history of BIM.

The popularity of Building Information Modelling (BIM) is obviously an exhibit A to showcase the benefits the software can bring to the construction industry. The ease of use in the software is an utmost important issue among busy construction practitioners. Besides this, the relative low cost is another important benefit that brought on by the software. Seniors in a construction company never dare to try something new if it costs them an arm and a leg, and this is particularly true when the construction company is having a trying time.

2 Software for BIM

2.1 BIM Software: Autodesk

If any man in the street can master software straight off and find it useful, that software will gain popularity easily. One of the very good examples of this is the drag and drop Autodesk BIM software (Figs. 2 and 3).

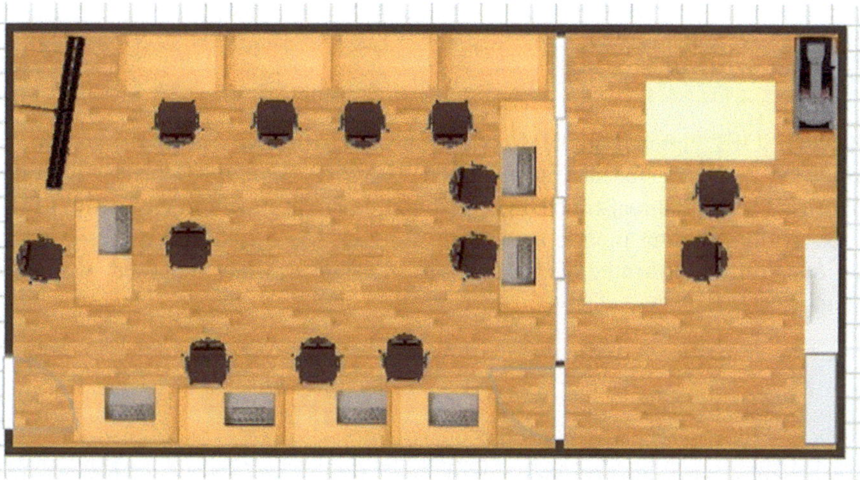

Fig. 2 Example of 2D floor plan for an academic laboratory by using Autodesk software (author's diagram)

Fig. 3 Example of 3D floor plan for an academic real estate laboratory by using Autodesk software (author's diagram)

2.2 Graphisoft

GRAPHISOFT® ignited the BIM revolution with ARCHICAD® in 1984. It was the first BIM software for architects. GRAPHISOFT developed the revolutionary BIMcloud®, the first real-time BIM collaboration environment in the world. BIMx® is the world's leading mobile application for BIM visualisation. EcoDesigner™ is the first fully BIM-integrated "GREEN" design solution released worldwide (Graphisoft 2016).

2.3 Planner 5D

People can design offices, houses, flats, cafes, bars, country-houses, dream houses or anything in two- or three-dimensional models via the apps for iOS, Android or OS X or the tool online. The Planner 5D online tool can assist us in creating home plans and modern interior design which resemble a professional designer in the absence of any professional skills. Figure 4 shows the proposed facilities in Real Estate and Economics Lab under BIM. Figures 5, 6 and 7 show the real Lab.

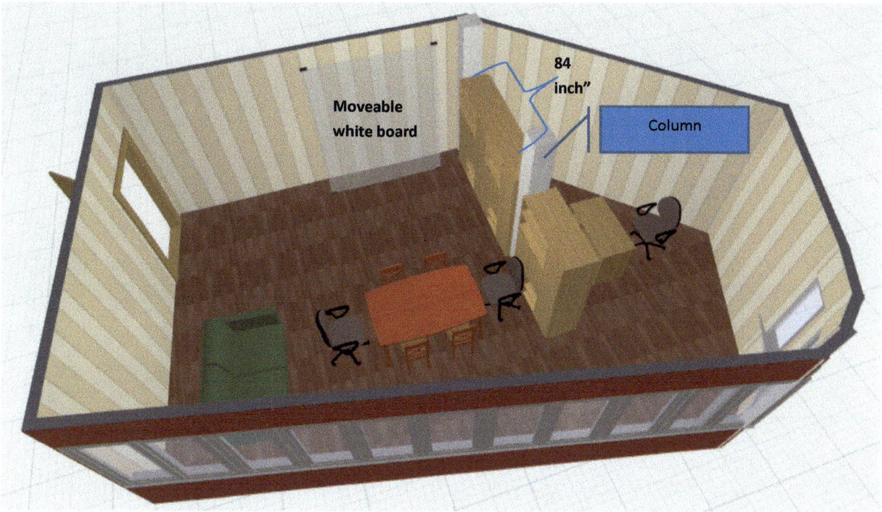

Fig. 4 Real estate and economics research lab's plan under the Planner 5D (author's figure)

Fig. 5 Real estate and economics research lab (author's figure)

Figure 8 illustrates the proposed Sustainable Real Estate Research Centre under BIM. Figures 9 and 10 display the real research centre.

The figures reflect some major drawbacks in the free edition of the software: users can only include the square column in the free one (Fig. 4), the circle ones can only be found in the paid version. In addition, the software does not allow us to include the three-dimensional staircases in the design (Fig. 8) which should be there in reality as shown in Fig. 9.

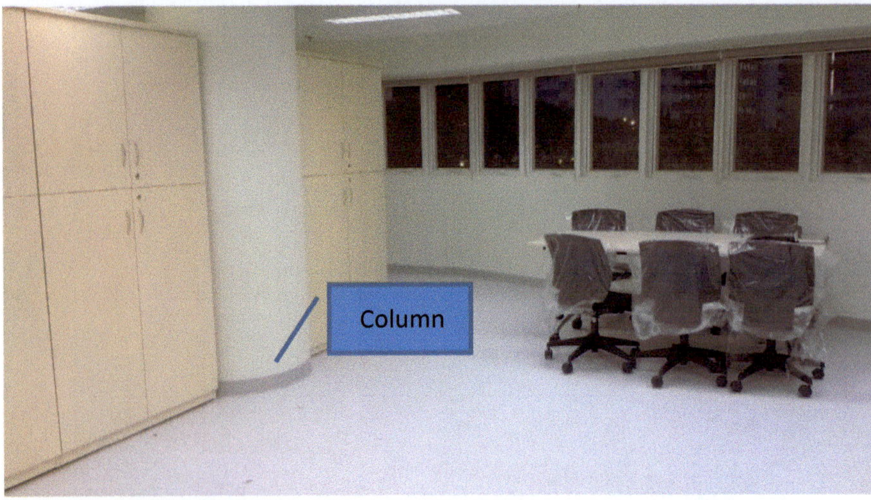

Fig. 6 Real estate and economics research lab (author's figure)

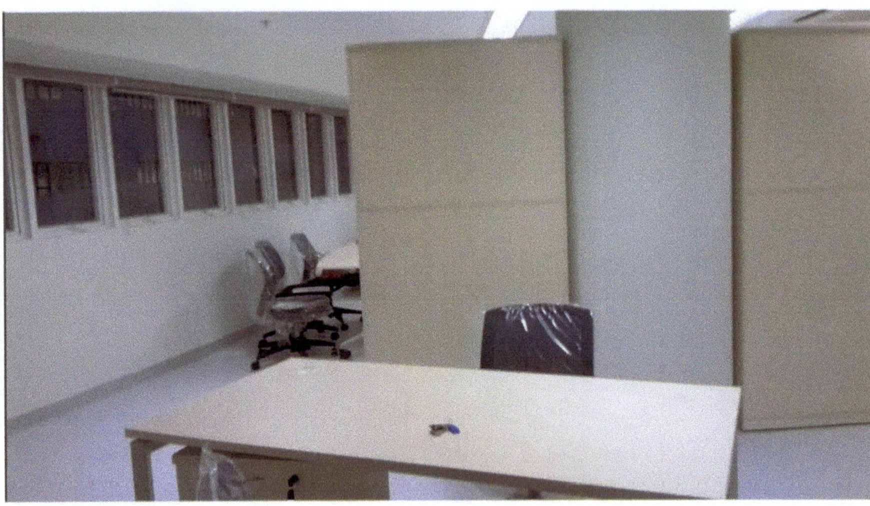

Fig. 7 Real estate and economics research lab (author's figure)

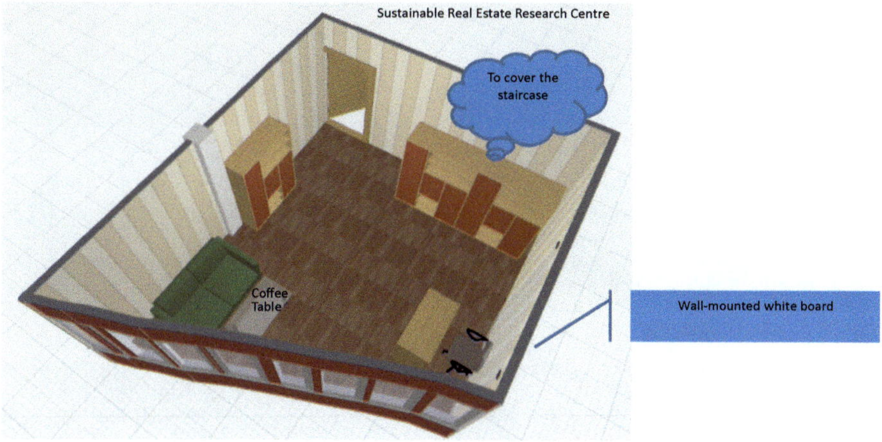

Fig. 8 Proposed Sustainable Real Estate Research Centre under Planner 5D (author's figure)

Fig. 9 The Sustainable Real Estate Research Centre in real life (author's photo)

Fig. 10 The Sustainable Real Estate Research Centre in real life (author's photo)

2.4 *UE4*

In UE4, a database is created with a project reference ID on the server, and the material is assigned for particular IFC classes. Windows, for instance, may be classified under the category of glass. UE4 materials can be retrieved by means of their path in the project directory. Besides this, the user can leverage default settings and override them via specific settings by adding a supplementary mapping set and using a project template (Hilfert and König 2016).

Due to the free Unreal Engine 4's flexible plugin, it can connect to different systems. A central open source server which holds all the BIM data is beneficial when the workers work in a team. First, an architect or engineer creates BIM models through a regular design process that can import their design files to the BIMServer, which can then parse the geometry and store it as BIM model data. The project manager may then manage different revisions in the building development process and update the data when the construction process takes place. The project is imported to the UE4 editor for further modification, bundled and then distributed to different machines for testing. As UE4 does not know about different BIM materials, the person in charge has to assign different materials. The UE4 code interfaces with the OpenSource BIMServer are exchangeable with different Application Programming Interfaces (API), and it is possible to implement custom protocols (Hilfert and König 2016). Figure 11 illustrates the UE4 BIM system and Fig. 12 shows the example of immersive testing of escape route visibility in UE4.

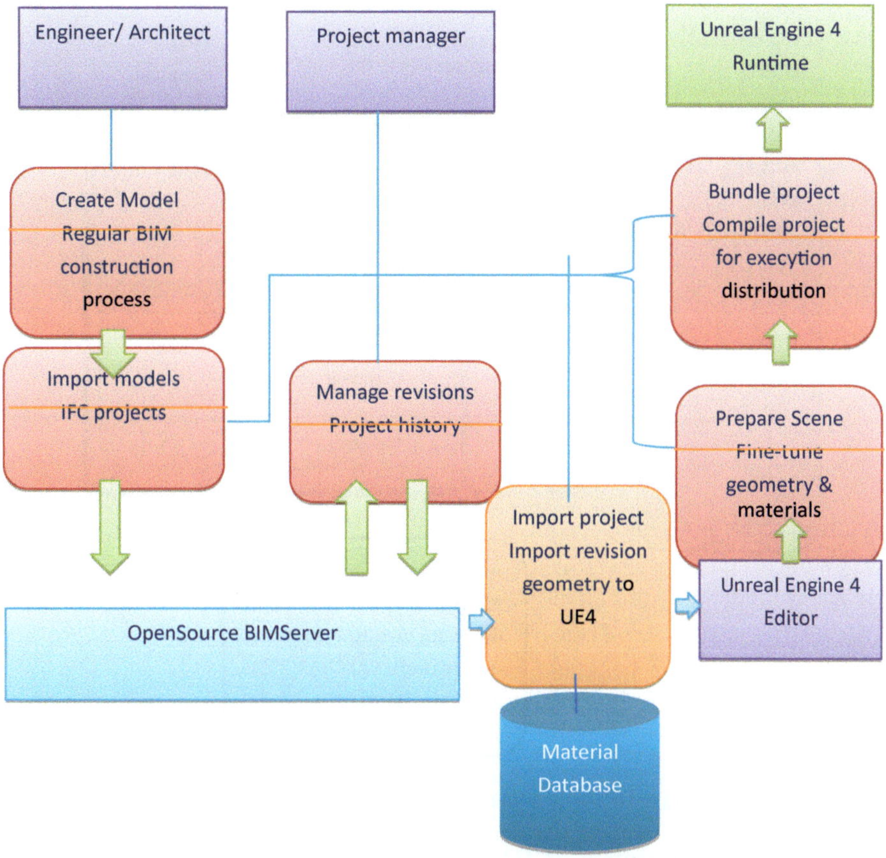

Fig. 11 UE4 BIM system (Hilfert and König 2016)

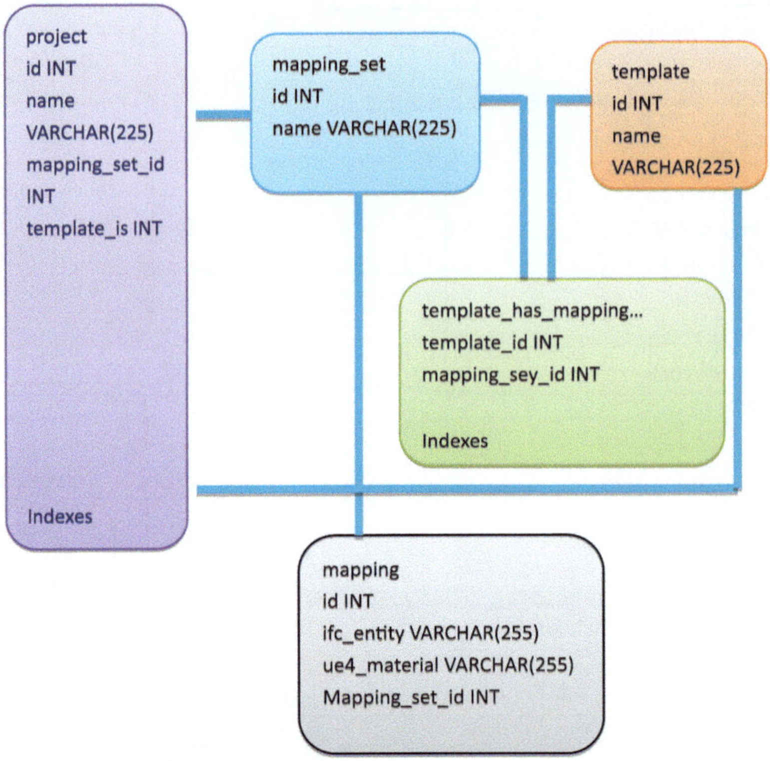

Fig. 12 Immersive testing of escape route visibility in UE4 (Hilfert and König 2016)

3 Costs and Benefits of BIM

Many of the previous literatures suggest the benefits that can be brought by information technology such as Web 2.0 (Li 2010; Li and Poon 2011), BIM is no exception. Early building information capture, integrated procurement, reduced conflicts during construction stage, improved cost control mechanisms and interoperability capabilities throughout the building life cycle, etc. are some of the benefits of BIM (Ghaffarianhoseini et al. 2017), and the above-mentioned features only represent an elementary level of it. A construction company may make adjustments to the software to enhance safety management.

Cho with Kim is the pioneer who explores the BIM-safety integration framework with the utilisation of rule-based safety-checking algorithms which automatically identify safety hazards in construction sites and notice the team members on site immediately. This information allows project managers and safety engineers to test the impact of different spatial workflows and scaffolding types on safety. It can also facilitate making preconstruction decisions that minimise hazards. Cho then explores the benefits of an iBeacon-based safety tracking system (Alderton 2015). Within a mobile BIM environment, when workers are near a potential safety hazard, they could receive real-time alerts on their mobile devices—reducing accidents by increasing awareness. In 2012, New York City became the first U.S. municipality to approve a three-dimensional Site Safety Plans Program by using BIM software, which allowed the construction industry to create and file the site safety plans electronically (Alderton 2015).

It is quite often that BIM is believed to be a catalyst for construction practitioners to improve and stimulate their safety concerns and their acknowledgement of the dynamic site environment while they are working in construction areas (Ganah and John 2015). In general, it alleviates the safety problem via better design, documentation and real-time visualisation.

3.1 Accidents Prevention via Better Design

The Bong et al. (2015) research shows that the designer plays a very important role in construction safety. BIM provides a powerful new platform for the advancement of "prevention through design" concepts to facilitate both engineering and administrative safety planning at the design and construction stage of a project (Zhang et al. 2013).

3.2 Benefits of BIM

3.2.1 Documentation Function in BIM

Building information modelling (BIM) is an object-oriented information-integration platform that provides a useful means for concurrent as-built documentation during the construction phase of a project. Construction records integrating with BIM remains a challenge due to unstructured data and heterogeneous formats. A WBS code-based link places the various construction project tasks together, and BIM objects can be created to generate the multidimensional BIM database (Park and Cai 2017).

BIM has been widely used in construction safety management research for the whole building life cycle of the modern construction industry. With regard to the safety management, BIM has been applied in the design stage to identify and assess safety risks on sites. Previous research has confirmed that the BIM-based augmented reality provides beneficial temporal information and aids safety training (Ding et al. 2013).

3.2.2 Real-Time Visualisation for Clash Detection

Clashes between different tasks on sites is one of the major causes of accidents (Ciribini et al. 2016). The BIM model allows the structure to be built virtually before the real ones are constructed, thereby allowing various construction stakeholders to detect clashes before the project is put into real practice. An optimal sequence can be arranged, which is impossible to do through using paper-based representations exclusively (Howard et al. 2017). Inevitably, clash detection is one of the most important characteristics of BIM during the project design stage. With the aid of clash detection, project designers or architects can check whether a proper work plan has fulfilled the relevant construction safety ordinance and ensure the best safety practice on site is implemented later (Ciribini et al. 2016).

The four-dimensional BIM presents a real-time visualisation safety status under changing conditions. Construction safety risks, workers, environment and structures can be monitored in real time, preventing accidents through the marked unsafe status of the environment and structures or the unsafe behaviour of workers (Ding et al. 2013). It also states the requirements for designers to validate the safety plan, including the period for installing scaffolding and formwork, and it records the time when the construction trucks will enter the sites. Thus, likelihood of having accidents on site will be reduced (Ciribini et al. 2016).

3.2.3 Identification of Safety Risks

Autodesk Revit, Microsoft Project and other application programming interfaces are used for construction safety. They show the design of the construction project well and let the project designers and architects simulate and eliminate any dangerous construction work prior to it being put into practice. Some BIM models simulate the temporary building structures, such as scaffolding, for an automated safety-checking approach. With the help of BIM and other software, all the relevant stakeholders can run the algorithm and see if there are any potential safety hazards in the early planning stages. Thus, appropriate actions can be followed (Kim et al. 2016) (Fig. 13).

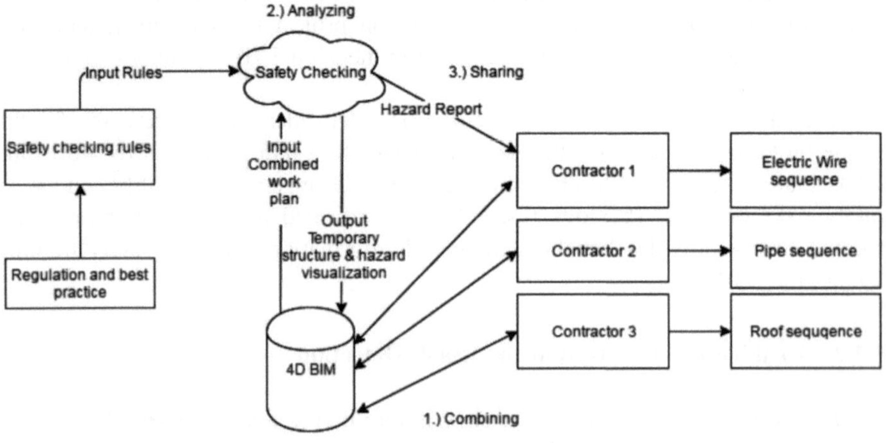

Fig. 13 Framework of automated BIM-safety platform (Kim et al. 2016)

Previous research shows that BIM is able to identify 40% of the potential risk in those construction projects. By using the software, the construction risks can then be lowered (Malekitabar et al. 2016). In order to use BIM for construction safety, there are eight different components which should be considered, including IT capacity, technology management, trust and others. BIM provides 3D models which visualise and display the clash detection. Hence, construction safety can be enhanced. With the aid of three-dimensional BIM, various stakeholders on site can detect any clashes and implement suitable action before construction works take place. Therefore, money and lives can be saved (Goh and Askar Ali 2016).

3.2.4 Flexibility in Combining Other Information Technology

Digital representation of the functional and physical characteristics of BIM allows users to transfer design data and specifications between different software (Ghaffarianhoseini et al. 2017). Sometimes, BIM can combine with other technology, such as ontology, and semantic web technology to identify the possible construction risks on sites. Previous research has adopted the semantic retrieval mechanism to link up with the objects in a BIM environment. A new system is created to investigate the possibilities of construction safety risk knowledge management. It identifies the potential risks in the construction process and analyses the risk factors and risk paths that are present (Ding et al. 2016).

3.2.5 Incorporation of Big Data in BIM to Enhance Construction Safety

Bilal et al. (2016) found out that the construction industry has started to use BIM with big data to analyse construction projects. As BIM can integrate with a large amount of different data it can provide a clear picture with regard to work assignments, reduce the error-prone procedures and replace the relative less effective site checks which should be originally done by building managers. It can also measure and present the live daily progress update to the manager-in-charge, which should make it easier for managers to control and improve the safety on sites.

3.2.6 Benefits on Management

It offers a range of potential benefits for the building environment, ranging from cost management in the planning and building stages, energy and retrofit management in occupation stage and historical monument management before the building is considered for demolition (Ghaffarianhoseini et al. 2017; Howard et al. 2017). In recent years, expenditures for real estate and facility represent the largest sum of the operating expenses for a company after wages. Any improvement in facility cost management results in a significant costs saving. The adoption of information and communication technologies (ICT) such as BIM has affected property management over the past few years. It provides an efficient application in controlling, mapping and evaluation of the facility management processes (Pak and Li 2012). The National Institute of Building Sciences suggests that BIM utilises cutting edge digital technology to establish a computable representation of physical and functional characteristics of facilities' life cycle information, and it is intended to be an information repository for the facility owner or operator to use throughout the life cycle of a facility (Howard et al. 2017) (Fig. 14).

Fig. 14 BIM's Benefits in different aspects of management (Ghaffarianhoseini et al. 2017)

Table 1 Price of various types of BIM software (Autodesk 2017)

Types of BIM software	Price	Functions
Revit	$2000/year	It delivers higher quality, accurate model-based architectural designs and building systems It makes better design decisions with more accurate visualising structures before they are built It facilitates streamlined collaboration by using information-rich models which ease the design decisions-makings It works with a single intelligent three-dimensional model to help improve coordination and collaboration across the project team regardless of where they are located It minimises coordination mistakes and reworks, improve how users communicate the appearance and intent of a design Documentation tools help users stay coordinated regardless of the number of changes in design It helps lower costly design conflicts during construction and minimise design coordination errors via interference detection It uses intelligent three-dimensional models and accurate documentation to enhance building systems designs in building life cycle
AutoCAD	$1470/year	It creates and shares precise drawings with innovative productivity tools It saves time and minimises frustration with simple tools to repair the broken paths for externally referenced files It uses the SHX text recognition tool to convert the imported PDF geometry quickly to text objects Experience significant improvements in 3D navigation when zooming and panning It utilises the power of AutoCAD everywhere with an easy-to-use mobile app
3ds Max	$1470/year	It improves editing and navigation with mapping enhancements feature. It addresses user's requests such as visual feedback and texture creation improvements It creates hard surface forms by using double precision Booleans which enable users to add and remove operands quickly, easily and reliably. It provides more inbuilt and well-organised tool access for object work Improvements to the Point to Point Selection tools, Sub Object Pick, conLocal Align and Working Pivot streamline the workflow It renders more near-photorealistic results for architectural projects using the Autodesk

(continued)

Table 1 (continued)

Types of BIM software	Price	Functions
		Raytracer renderer, a physical-based renderer which is ideal for other Autodesk applications' visualisation workflows
Fusion	$300/year	Design differently using a single tool which combines mechanical and industrial design with collaboration in an affordable and easy-to-use package It allows users to access the designs anytime and anywhere from a computer or mobile device. With the help from Autodesk Fusion 360 for 3D CAD and CAM users serve as an excellent integrated design tools by getting a complete set of mechanical and industrial design capabilities within one package: • Anywhere access: users can access their designs anytime and anywhere from mobile device or computer • Advanced collaboration: share the design information and start working together via the secure dashboard • Fabrication: prepare designs with an integrated CAM tool with 2.5 and 3 axis machining or utilise the three-dimensional printing capabilities for additive manufacturing • Animation: assemble components. Create animation and motion simulation, review the function of and show the relationship of various forms of design
Architecture Engineering and Construction Collection	$2690/year	It allows the users to access to a wide range of three-dimensional design software. It provides interoperable tools for CAD and BIM-based workflows It connects CAD-based processes for efficient documentation and design with the help of three-dimensional model-based design workflows It allows us to choose the individual products that we wish to use; the usage of mobile apps and the cloud enable us to view and create stunning visualisations
Product Design Collection	$2460/year	It allows us to access to a wide range of three-dimensional design software. It helps to improve productivity by using tools such as AutoCAD Mechanical or AutoCAD Electrical. It uses mobile apps and the cloud view to create rich visualisations and store the building structure designs

3.2.7 Economic Benefits: Return on Investment and Building Life Cycle Analysis (LCA) Benefits

Research has confirmed that BIM prevents delays and has the most influence on increasing the Return on Investment (ROI). It also prevents any unnecessary reworking based on the initial model validation (Ghaffarianhoseini et al. 2017). BIM is also advantageous throughout the building project life cycle from the early conceptual design stages to demolition activities. In the planning stage, since all the building materials and components are predetermined and their corresponding quantities are calculated automatically, building materials and components can be ordered electronically and delivered just in time, thereby increasing workers' productivity (Ghaffarianhoseini et al. 2017).

3.3 Costs of BIM Software

Using BIM is not without its costs, and some companies do not use it mainly due to pricing concerns. In general, the costs of BIM vary from US$300 per year to US $27,900 depending on the needs and requirements of the companies (Autodesk 2017). Table 1 shows the price of various types of BIM software (Autodesk 2017).

4 Research Method

In this chapter, we have included a case study on one of the largest construction companies in Hong Kong and one of the largest construction sites in Adelaide.

4.1 BIM's Application in Hong Kong

There is relatively long history with BIM compared to other automated tools such as three-dimensional printing and drones. Similar to the US,[1] it is not very popular in Hong Kong according to officers that we met in an architecture government office. They admit that most of the building projects in their departments have not yet applied BIM as the cost is high, and some of the officers have not learned about BIM before. Yet, they foresee that there should be an increase in popularity as the university and many other course providers include BIM in their syllabus. In view of this, it is high time to seize the opportunity to adopt BIM on site.

[1]According to the author's interview with some participants in 53rd Associated School of Construction in Seattle 2017.

In another large-scale private construction company in Hong Kong, BIM is applied to draft the building structure. The three-dimensional BIM structure is then transferred to a three-dimensional printable object. The model is also used to examine the possible building structure problem. The company admits that the original driven point of adoption of 3D printing was not used for improving safety but was used to satisfy the needs of the clients (Fig. 15).

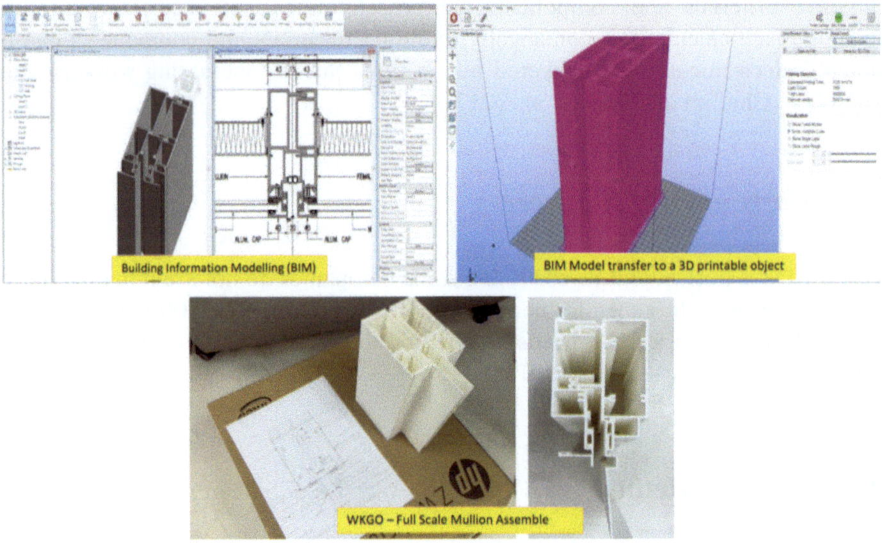

Fig. 15 BIM and 3D printed model (information provided by Hip Hing Construction Limited)

Table 2 Functions and characteristics of different dimensions of BIM

Number of dimensions of BIM	Functions and characteristics of BIM
Two	It shows the X–Y plan of architecture design
Three	It shows the X–Y–Z plan of the architecture design where the height of the building structures are included
Four	Time is included. Builders know which parts have to be built and thus reduce the likelihood of work clashes, which increases danger on the sites
Five	The facility's information is included; for example, date of the warranty expired, function of the tools, etc. facilitating the repairment later during the occupation stage

The BIM officer of this company suggests that this inevitably enhances construction safety on site as they may better foresee if there are any building structure problems before the building is constructed, allowing them to revise the design before there any problems take hold. Figure 11 illustrates the BIM and the three-dimensional printed model which is printed according to the BIM design.

On the other hand, an innovation manager from one of the largest private construction companies commented that their company had already incorporated more than four-dimensional BIM in their projects. He recalled that one project in his company had included facility management according to the needs of the clients. As the facility manager always has difficulty in identifying the utilities, conduits and other facility when they need to be repaired, the BIM which is used in the construction stage shall benefit the facility manager, who can rely on it. Hence, apart from the traditional three-dimensional X–Y–Z coordinate design and time dimension, facility management has been incorporated. He also mentioned that there were varieties of different forms of BIM with different dimensions incorporated. Many of these appeared due to the clients' requests. Table 2 shows the functions and characteristics of different dimensions of BIM which are applied in the construction industry.

4.2 *BIM's Application with Internet of Things in a Hospital Building Project in Adelaide*

BIM has been applied in a hospital project in Adelaide to visualise the building process before the buildings are built. The system provides useful information with regards to what materials will be used, how much, what to order and where to instal so as to reduce waste and work more efficiently. They can also utilise the BIM system together with team viewer (another piece of software). Such that construction teams in Melbourne and Brisbane can collaborate with the teams in Adelaide. As they can outsource part of the work to other construction teams outside Adelaide, everyone can do one's bits according to their strengths to make the best of the job. An information technology officer in the Adelaide Hospital suggests that, "*For example, if they do not know how to design three-dimensional model, they can outsource this part, and the team in Adelaide can use the three-dimensional model designed by other team for analysis. To a certain extent, BIM opens up the opportunity for international collaborations, as teammates may come from everywhere around the globe, from China to Peru*".

On top of that, they also incorporated the time dimension in the BIM model when they designed the building facade, and the timeliner shows how things should be installed when all the jobs have been completed. It also allows them to foresee if there is one particular time where they are too many people on site, causing safety hazards. For example, a concrete truck is planned to enter the site at 1:30 as scheduled, so the BIM system will inform the security guards so as to avoid many trucks coming up at the same time. BIM can also let us foresee how the room is built so that we will have a better understanding of which parts of the room should be built first, when we should lay the pipe down and make sure that it is in the correct sequence.

As objects falling from a height is one of the common causes of accidents on sites, BIM with timeliner can let us foresee when will we instal the window frames or scaffoldings and warn workers to leave before they are installed. All these can prevent possible hazards at work. That can also let the site manager be aware of what they have to do and ensure better coordination and communication between each other.

Besides this, every single piece of material used is not only stored in the RFID tag but also visualised in BIM. The BIM system is interconnected with offsite prefabrication system through the Internet of Things system. The system will alert them when everything is done. The system is stored in the cloud and backs up the project to avoid any problem happening (author's interview).

4.3 BIM's Application in Casa Magayon in Costa Rica

Casa Magayon in Costa Rica was modelled by ARCHICAD from its initial concept. The two-dimensional document was generated from the ARCHICAD model entirely in the absence of any two-dimensional diagram. Teamwork was enhanced throughout the development process, and BIMx was used as a communication tool with the client during construction phrase. ARCHICAD's MEP modeller tool enhanced the integration of plant trajectories, planning from the design phrase and analysis, offering integral solutions to the HVAC component, furniture, ceiling design and so on. Table 3 shows the background information of the BIM's application in Casa Magayon in Costa Rica. Figures 16, 17, 19 and 20 illustrate the design under BIM and the real building structure (Figs. 18 and 21).

With the help of an automated quantification, the architect generated several lists and schemes of zones, structures, elements, foundation and components, which were then integrated as an essential part of the document in the project. The entire

Table 3 Background information of the BIM's application in Casa Magayon in Costa Rica (Graphisoft 2016)

Project name	Casa Magayon
Location	Guanacaste, Costa
Type	Residential
Award type	Central and South America Property Award for Casa Magayon
Architect	Roderick Anderson
Year	2015
Size	13.900 ft^2/1290 m^2
Software used	GRAPHISOFT ARCHICADGRAPHISOFT BIMx

Fig. 16 Three-dimensional plan of the housing BIM is created with GRAPHISOFT ARCHICAD

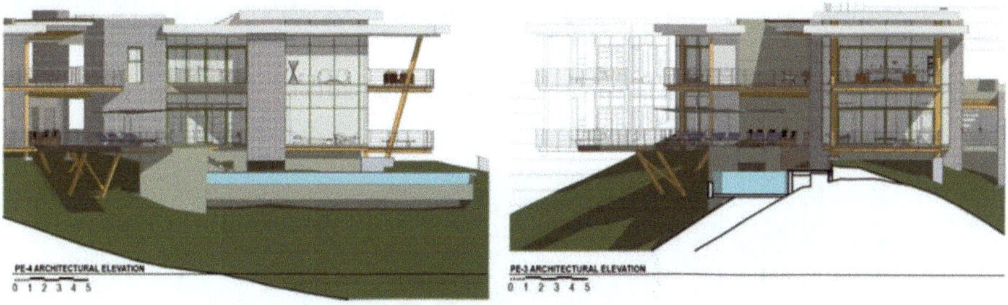

Fig. 17 Architectural elevation BIM is created with GRAPHISOFT ARCHICAD (Graphisoft 2016)

project was generated by "Viewmap" and "layer combinations", which allowed them to save views and layers that are passed to documentation in absence of the needs to have different BIM models for the different stages of work. All the quantification of carpentry, elements and materials could be handled automatically, with all the necessary technical sheets with its suppliers, allowing the project to finish more efficiently, flawless and a lot faster.

"The results of working with ARCHICAD as a BIM tool are surprising, since throughout the life cycle of a project we can accompany it in its different stages, witnessing that the model is identical to reality as our project begins to take shape" CEO & Design Director of SARCO Architects Costa Rica comments (Graphisoft 2016). The BIMx application opened up the possibility of combining two-dimensional content with three-dimensional content on mobile devices, altering the logic of a plan's graphical representation. The application showed the customer the level of precision during preliminary presentations. Colour was incorporated in the final

Fig. 18 Construction of the building in progress (Graphisoft 2016)

Fig. 19 Casa Magayon BIM is created with GRAPHISOFT ARCHICAD (Graphisoft 2016)

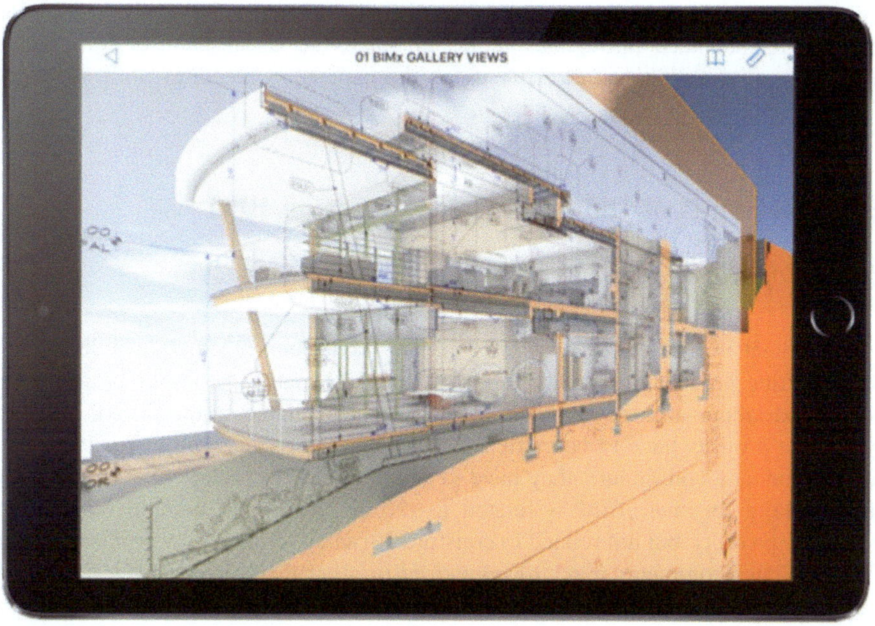

Fig. 20 Design shown in mobile apps (Graphisoft 2016). BIM is created with GRAPHISOFT ARCHICAD

Fig. 21 The BIM model shown in the mobile apps (Graphisoft 2016)

documentation. Hence, BIMx became a central communication and visualisation element for architects throughout the design stage, the visualisation of complicated projects, and as an essential element for discussions in meetings with suppliers, architects and so on (Figs. 20 and 21).

5 Viewpoints of Different Stakeholders on BIM

5.1 Popularity of BIM in Recent Years

An architect in Hong Kong suggests that "BIM in the construction industry has been getting prevalent gradually. The costs of BIM mainly lie on the employment of skilled labour and the software (if that is not available in the firms). Besides, when the projects have to be carried out by various different parties including the contractors and consultants, they have to acquire the relevant knowledge and skills for working together. BIM is related to the efficiency and complexity of construction projects, but not related to construction safety directly". A professor in the United States comments that, "*as compared to other technologies such as robots, regulations in BIM is well developed. Companies are willing to try BIM. In the United States, we have the most advance in BIM technologies*".

5.2 Costs and Benefits of BIM

A CEO of a tool provider in Guangzhou suggests that "the benefit is smarter design, the two-dimensional building plans are rapidly being replaced by three-dimensional computer models, so people can preview the complicate design. The cost is need to design hardware, software and research". A section chief in Fujian China comments that, "It is convenient. However, the information obtained from BIM is not well connected and shared among relevant stakeholders".

A researcher in Melbourne suggested that "The costs include hiring BIM management company to undertake the project. The benefits are by using BIM, the company can evaluate construction process and equipment operations, then determine safety hazards to workers and engage in the appropriate measures before accidents occur and reduce the accident rate".

In the United States, BIM fasten the pace of robotic development in the construction industry due to the relative high requirements in law. The latest development of BIM in the United States is to combine BIM with robots, such that the robots work according to the requirements stated in BIM. As it is quite difficulty to overcome the legal barrier of adopting the unproven technology such as robots on sites, BIM provides them a very good explanation to the legal authority with regard to robots to be used on sites "*BIM is now lowering the legal barriers of robots. In the United States, we are trying to incorporate robotics in the context of BIM*".

In addition, he pinpoints the impact of different types of procurements on the potential benefits of BIM "*Most public works are design-bid built projects, designers do not know who the builders will be. On the other hand, in case of Design build, when contractors bid for the projects do the projects, you know how easy or difficult to build at the design stage, BIM allows all parties include the suppliers, contractors and even the general public access to the documentation. But when the contractors are not selected in design-bid built projects, the contractors have no participation in the building model, including the BIM*".

6 Conclusion

BIM has been developing at full tilt in recent years. Some construction companies have already blazed a trail in combining BIM and 3D printing to better foresee future difficulties and problems in construction work. Despite one criticising that it happens once in a blue moon and may be considered as a back burner only, as these technologies can save the costs and do not burn the candle at both ends, we may foresee that the benefits brought by them will turn that popularity across the board sooner or later. More importantly, whilst information technology officers are seen as a cog in the wheel in the construction industry, BIM may convince senior construction companies to hire them for problem solving in various areas. Whilst costs saving may be one of the major issues, it is on cue for shedding light on safety issues as well.

Acknowledgements The author would like to thank Miss Siu Han, Mary and many other research assistants' contribution in this chapter.

References

Alderton, Matt. 2015. Modeling safety: BIM and construction research may help prevent injuries and save lives. In *Construction*: Line//Shape//Space.

Autodesk. 2017. *Autocad Lt* [cited 24 Apr 2017]. Available from http://www.autodesk.com/store.

Bilal, Muhammad, Lukumon O. Oyedele, Junaid Qadir, Kamran Munir, Saheed O. Ajayi, Olugbenga O. Akinade, Hakeem A. Owolabi, Hafiz A. Alaka, and Maruf Pasha. 2016. Big Data in the construction industry: A review of present status, opportunities, and future trends. *Advanced Engineering Informatics* 30 (3): 500–521. doi:10.1016/j.aei.2016.07.001.

Bong, Shanan, Raufdeen Rameezdeen, Jian Zuo, Rita Yi Man Li, and Gui Ye. 2015. The designer's role in workplace health and safety in construction industry: Post-harmonized regulations in South Australia. *International Journal of Construction Management* 15 (4): 276–287.

Ciribini, A.L.C., S. Mastrolembo Ventura, and M. Paneroni. 2016. Implementation of an interoperable process to optimise design and construction phases of a residential building: A BIM Pilot Project. *Automation in Construction*. doi:10.1016/j.autcon.2016.03.005.

Ding, L.Y., C. Zhou, Q.X. Deng, H.B. Luo, X.W. Ye, Y.Q. Ni, and P. Guo. 2013. Real-time safety early warning system for cross passage construction in Yangtze Riverbed Metro Tunnel based on the internet of things. *Automation in Construction* 36 (25-37).

Ding, L.Y., B.T. Zhong, S. Wu, and H.B. Luo. 2016. Construction risk knowledge management in BIM using ontology and semantic web technology. *Safety Science* 87: 202–213. doi:10.1016/j.ssci.2016.04.008.

Ganah, Abdulkadir, and Godfaurd A. John. 2015. Integrating building information modeling and health and safety for onsite construction. *Safety and Health at Work* 6 (1): 39–45. doi:10.1016/j.shaw.2014.10.002.

Ghaffarianhoseini, Ali, John Tookey, Amirhosein Ghaffarianhoseini, Nicola Naismith, Salman Azhar, Olia Efimova, and Kaamran Raahemifar. 2017. Building Information Modelling (BIM) uptake: Clear benefits, understanding its implementation, risks and challenges. *Renewable and Sustainable Energy Reviews* 75: 1046–1053. doi:10.1016/j.rser.2016.11.083.

Goh, Yang Miang, and Mohamed Jawad Askar Ali. 2016. A hybrid simulation approach for integrating safety behavior into construction planning: An earthmoving case study. *Accident Analysis & Prevention* 93: 310–318. doi:10.1016/j.aap.2015.09.015.

Graphisoft. 2016. *A complete BIM residential project SARCO Architects Costa Rica wins Central & South America Property Award for Casa Magayon* [cited 4 Feb 2017]. Available from http://www.graphisoft.com.hk/users/bim-case-studies/sarco-architects-magayon-house.html.

Hilfert, Thomas, and Markus König. 2016. Low-cost virtual reality environment for engineering and construction. *Visualization in Engineering* 4 (2): 1–18.

Howard, Robert, Luis Restrepo, and Chen-Yu. Chang. 2017. Addressing individual perceptions: An application of the unified theory of acceptance and use of technology to building information modelling. *International Journal of Project Management* 35 (2): 107–120.

Kim, Kyungki, Yong Cho, and Sijie Zhang. 2016. Integrating work sequences and temporary structures into safety planning: Automated scaffolding-related safety hazard identification and prevention in BIM. *Automation in Construction* 70: 128–142. doi:10.1016/j.autcon.2016.06.012.

Li, Rita Yi Man. 2010. Knowledge sharing by Web 2.0 in real estate and construction discipline. In *E-Research collaboration: frameworks, tools and techniques*, ed. Murugan Anandarajan, and Asokan Anandarajan, 289–301. Berlin, Heidelberg: Springer.

Li, Rita Yi Man, and Sun Wah Poon. 2011. Using Web 2.0 to share the knowledge of construction safety as a public good in nature among researchers: The fable of economic animals. *Economic Affair* 31 (1): 73–79.

Malekitabar, Hassan, Abdollah Ardeshir, Mohammad Hassan Sebt, and Rudi Stouffs. 2016. Construction safety risk drivers: A BIM approach. *Safety Science* 82: 445–455. doi:10.1016/j.ssci.2015.11.002.

Pak, Donald Henry Ah, and Rita Yi Man Li. 2012. Strategically managing facility management knowledge sharing via Web 2.0. *Computer Applications for Communication, Networking, and Digital Contents. Communications in Computer and Information Science* 350: 260–264. doi:10.1007/978-3-642-35594-3_36.

Park, Jaehyun, and Hubo Cai. 2017. WBS-based dynamic multi-dimensional BIM database for total construction as-built documentation. *Automation in Construction* 77: 15–23.

Zhang, Sijie, Jin-Kook Lee, Manu Venugopal, Jochen Teizer, and Charles Eastman. 2013. Integrating BIM and safety: An automated rule-based checking system for safety planning and simulation.

Chapter 4
Addictive Manufacturing, Prosumption and Construction Safety

Abstract Construction safety is one of the most important issues on-sites. In recent years, as the costs of construction accident compensation increased, there has also been a growth in the number of the research studies that shed light on the possible solution to lower the likelihood of accidents. Whilst many of the previous research showed that modular construction reduces the construction accident rates due to higher level of certainty in factory-like environment, we conjecture that the new additive manufacturing (AM) technology may serve a similar purpose. We wish to explore the possibility of using additive manufacturing to reduce construction accidents on-sites. We first provide a brief introduction about additive manufacturing. After that, we provide a detailed description about the working principle of additive manufacturing. Third, we study various different factors that affect the quality of additive manufacturing. Fourth, the pros and cons of additive manufacturing in the construction industry will be explained. Additionally, the application of additive manufacturing in the construction industries will be illustrated. Last, but not least, some case studies will be included to illustrate the application of additive manufacturing in the construction industry to ensure safety on-sites.

Keywords Additive manufacturing · 3D printing · Cost and Benefit analysis · Construction safety

1 Introduction

The sparkling glass buildings' facade, speedy construction of skyscrapers and a close-to-ideal blend of city showcase human's collective wisdom. They also evidence the sweats and tears of engineers and architects. Nevertheless, signatures of human civilisations as such are swathed in towering statistics of construction

R.Y.M. Li, *An Economic Analysis on Automated Construction Safety*,
DOI 10.1007/978-981-10-5771-7_4

accidents (Li and Poon 2013a, b). It is one of the most high risk industries beyond question (Li and Poon 2011, 2013a). According to the Occupational Safety and Health Statistics Bulletin Issue No. 15 (2015), the construction industry in Hong Kong alone recorded 3467 industrial accidents in 2014 which accounted for 9.2% of the total number of occupational injuries in all workplaces (Li and Poon 2013a).

Many construction accidents occur as a result of a number of distant and direct causes (Li and Poon 2013a). The direct ones are the causes that have an immediate effect on workers' safety condition, such as slips on some materials and falls in days of heavy rainfall. The indirect ones include poor construction safety management, and lack of education, knowledge management and training. As most of the construction sites are not as clean as they should be, poor housekeeping on-sites is another major cause that leads to accidents. Examples of direct and indirect causes of accidents such as poor organisation and economic concerns are shown in Fig. 1.

To address these issues, different countries have adopted different safety measures for construction sites (Li and Poon 2013a, b). As many of the construction accidents on-sites are caused by bad weather, off-site constructions through precast methods have also been adopted as alternatives to traditional on-site construction processes over the past few years (Pan and Sidwell 2011). In modular construction, 85–90% of the work is completed in factory-like environment and then transported to the final project location to form a building. Different research studies reported various benefits from using off-site construction approach. For example, it reduces the level of elevated work and dangerous activities on-sites, reduces on-site workforce congestion, less numbers of workers are exposed to neighbouring construction operations, and less time is needed to be spent working on-site in severe weather. Others have found that it enhances construction sustainability (Kamali and Hewage 2017).

In modern days, apart from traditional precast/modular construction, another form of off-site construction is additive manufacturing (AM). However, the effectiveness of this approach has raised questions among academia and construction practitioners in different parts of the World as it is still in its infancy. While some of us may question the structure safety and other issues such as quality of the 3D printed house, there are an increasing number of eye-opening additive manufactured houses examples being shown in YouTube and Facebook.

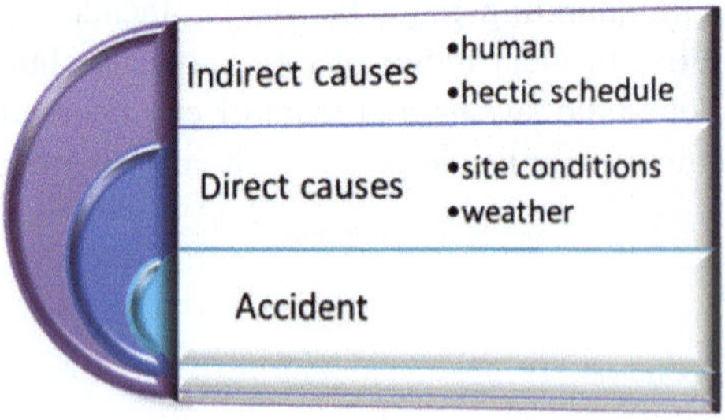

Fig. 1 Causes of construction accidents (Li and Poon 2013a, b)

2 Additive Manufacturing

Additive manufacturing (AM), also referred to as three-dimensional printing (3D printing), layered manufacturing (LM), rapid prototyping (RP), is a fundamentally different fabrication process from traditional manufacturing methods, which is performed by integrating material science, computer-aided design (CAD) and computer programming techniques to fabricate physical prototypes from virtual CAD models layer-by-layer (Jin et al. 2017). It cuts across the boundaries of many different disciplines such as material science, computer programming and construction.

The recent decades have witnessed the rapid development of AM technologies (Griffey 2014). It has experienced an impressive double-digit growth over the past 17 years (Pei et al. 2015), from something considered as an illusionary discovery, to gradually bringing it online for a wide range of everyday life applications. Some vivid examples include selective laser sintering (SLS), stereo lithography, laminated object manufacturing, electron beam melting (Griffey 2014), fast construction of millimetre-thick macroscopic tissues (Matsunaga et al. 2011), electronic devices, product development, biomedical engineering, prototype fabrication, architecture and so on (Jin et al. 2017) (Fig. 2).

The increase in popularity of AM has led to a stronger demand for entry-level three-dimensional printers (EL3DPs) and the low-cost desktop addictive manufacturing systems are proliferating in the prosumption market. Many of these are produced with the help of fused deposition modelling (FDM) which uses the extrusion of molten thermoplastics. Other processes including stereo lithography apparatus, selective heat sintering and SLS, digital light processing, etc., are increasing their market share (Pei et al. 2015).

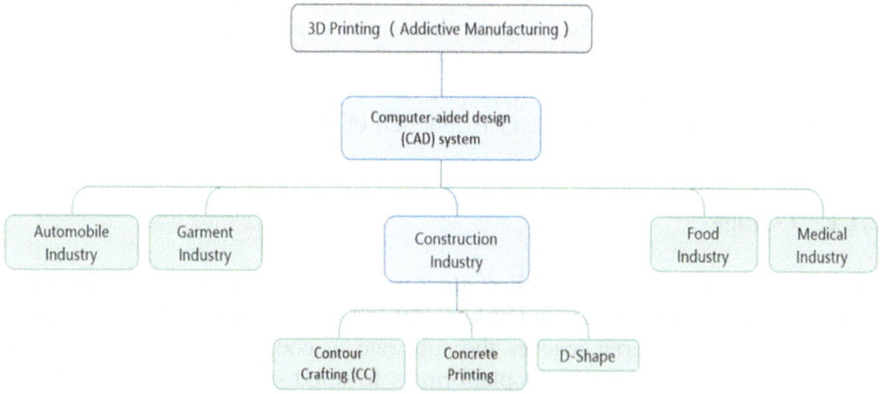

Fig. 2 Applications of additive manufacturing in different industries (Lee 2010)

Table 1 Filament materials and their properties (Pei et al. 2015)

Properties	ABS	PLA	Nylon 645
Chemical name/blend	Acrylonitrile butadiene styrene	Polylactic acid	Copolymer of nylon 6/9, 6 and 6T
Supplier	Form Futura	Ultimaker	Taulman
Cost (€/kg)	32.95	31.50	39.95
Diameter (mm)	2.85	2.85	2.85
Opacity	Opaque	Opaque	Transparent
Density (kg/m^3)	1010	1240	930
Print temperature (°C)	200–250	180–220	230–265
Melting temperature (°C)	105	65	194
Impact strength (KJ/m^2)	16	7.5	Not applicable
Tensile modulus (Mpa)	2000	3310	114

One major reason for the growing popularity is that key patents for FDM technologies has just expired and the open-source movement aligned with Raspberry Pi and Arduino microcontrollers is supported with free license (Pei et al. 2015).

Table 1 illustrates some of the common filament materials and their properties.

2.1 *Principles of Additive Manufacturing*

To realise AM, a three-dimensional model of the object needs to be designed using the CAD software or scanning the existing artefact (CAD will be further discussed in the following part). Specialised software slices the model into cross-sectional layers and creates a file in the computer that is sent to the AM machine. Finally, the AM machine creates the object by forming each layer through the selective placement of material. Different types of AM processes are differentiated by the approach taken to create each layer (Campbell et al. 2011).

2.1.1 Z-Stage

The AM is guided by the feed roller into a liquefier that is heated to a temperature above the AM agents' melting point first. After that, the material flows through the nozzle. Once the material reaches the substrate, it cools and hardens. When the layering is completed, the build platform is lowered one layer-thickness by the Z-stage and deposition of the next layer begins. A secondary supporting material is occasionally needed to support the construction of overhanging geometries (Fig. 3) (Campbell et al. 2011).

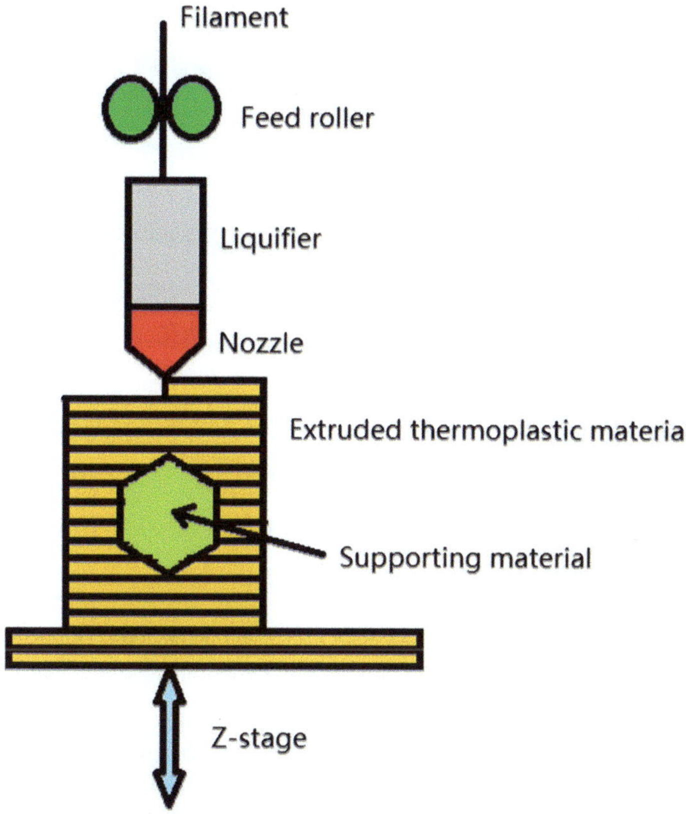

Fig. 3 The additive manufacturing process (Campbell et al. 2011)

2.1.2 Computer-Aided Design System (CAD)

As noted, the process of AM is supported by the CAD system, which can be defined as the use of a computer to create, modify, analyse, or optimise a design (Narayan et al. 2008). CAD systems can be applied to many different areas. Manocher (2012) suggested that the CAD system could be applied in making quantitative analysis for environment impact. The software simulates the environment of the intended building to find out the impact on the surrounding situation such as the carbon emissions, air pollution, energy and so on. In addition, CAD can be used in engineering.

Lee (2010) indicated that CAD could be used by product designers to make products for their customers. Wimmer et al. (2015) found that AM can also be achieved by CAD software, or three-dimensional scanning of an existing object. With the help from a slicing software, the three-dimensional objects can be separated into cross-sectional layers, resulting in a suitable computer file which can be transferred to a three-dimensional printer.

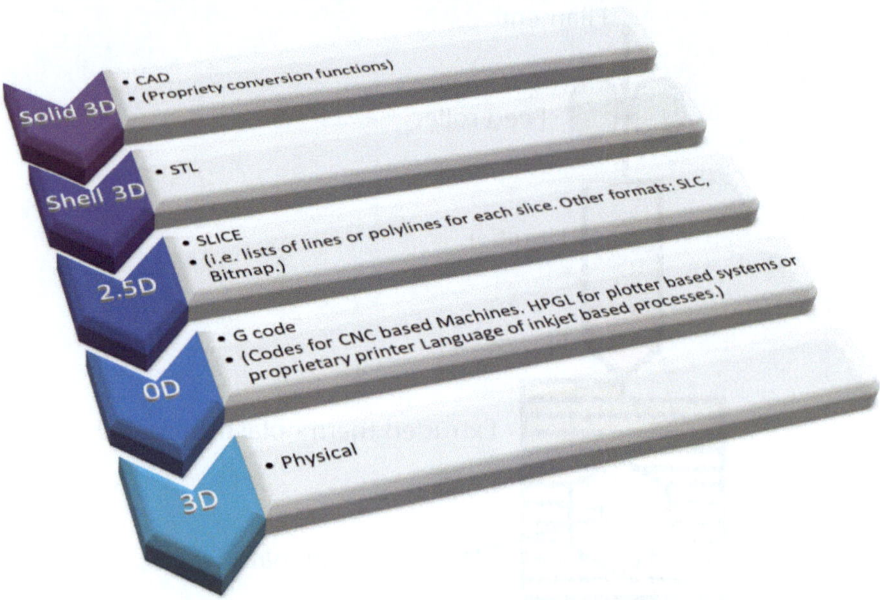

Fig. 4 Steps in producing an artefact using rapid manufacturing processes (Buswell et al. 2008)

Figure 4 depicts how to produce a physical object using AM concept. It must be a three-dimensional solid either using Constructive Solid Geometry (CSG) or Boundary Representation (B-rep). The modeller is a manifold modeller ensuring that the model can be sliced. Standard functions can be used to generate a STL file and then processed by Rapid Manufacturing machine-dependent software. They are then sliced and represented as a series of two-dimensional layers. The data are exported to a printing machine to reconstructs a three-dimensional object by sequentially bonding these '2D' layers of material (Buswell et al. 2007).

2.1.3 Contour Crafting

Contour crafting is a layered fabrication technology. It is defined as "an additive fabrication technology that uses computer control to exploit the superior surface-forming capability of trowelling to create smooth and accurate planar and free-form surfaces" (Khoshnevis 2004). Lim et al. (2012) indicated that contour crafting was based on extruding a cement-based paste against a trowel to make a smooth surface. In general, there are several advantages for contour crafting. For AM, Hwang and Khoshnevis (2005) suggested that CC had a strong potential in the construction area due to its low cost, simplicity, and ease to link to the current available technologies. Besides, contour crafting has the ability to produce overhangs by compression and avoid the cantilever problem. For example, AM, when a doorway or window is required a lintel is placed to bridge the gap and the wall can be placed above.

2.2 Factors Which Affect Quality of Additive Manufacturing

Chen et al. (2016) found that three major factors affect the final three-dimensional products: laser power, laser scanning speed and resin, and operating temperature. Tuteski et al. (2015) concluded that it was critical to concentrate on the thread size when make AM, as more layer printed often lead to better representation of the functionality of the final product.

3 Prosumption and Additive Manufacturing

In recent years, a substantial reduction in the price of three-dimensional printers for AM has led to a gain in popularity of three-dimensional printing activities. Many industries have realised the convenience and possible cost saving linked to addictive manufacturing, which has allowed the application of this technique to flourish. AM printing allows very low volume production at low cost and makes large-scale mass-customisation possible. It also creates the opportunity for co-creation between customers and firms, which is important in innovative production (Bogers et al. 2015).

With the help of AM techniques, co-creation takes place at different production stages: design, manufacturing and distribution stages. It also takes place between individual customers, leading to co-design communities. Similarly, mass-customisation is often associated with the production of tailored goods or services on a large-scale production. This increase in participation has flourished following the increasing popularity of Web 2.0 technologies, such as Instagram, Facebook, Twitter and Flickr. This rise in user participation blurs the line between production and consumption activities, with the consumer becoming a prosumer. The idea of prosumption, i.e. that production and consumption takes place at the same time, is likely to inevitably become one of the major industry trends in the near future (Bogers et al. 2015).

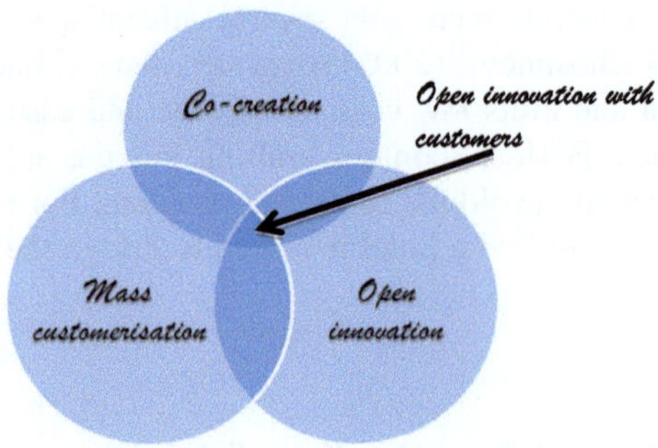

Fig. 5 Relationship between open innovation, mass customerisation and co-creation (Bogers et al. 2015)

Crowdsourcing entails contribution from users within the observing crowd, such as social media user site and followers. Online AM platforms allows a wide range of user involvement in the production process, from the lowest degree of participation, i.e. the user buys his own design that is printed and delivered by the platform to the user, who co-designs the object and prints at home (Bogers et al. 2015) (Figs. 5, 6 and 7).

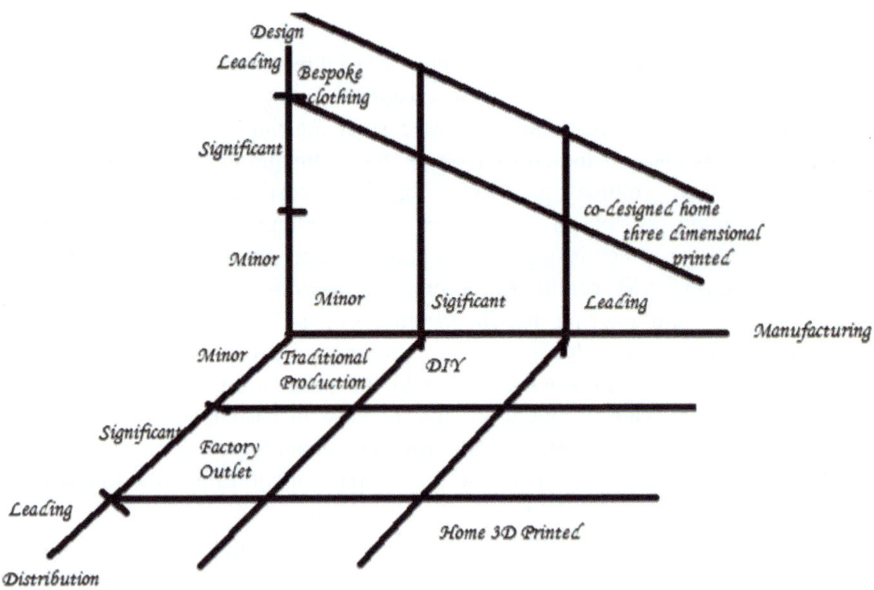

Fig. 6 Relationship between distribution, manufacturing and design (Bogers et al. 2015)

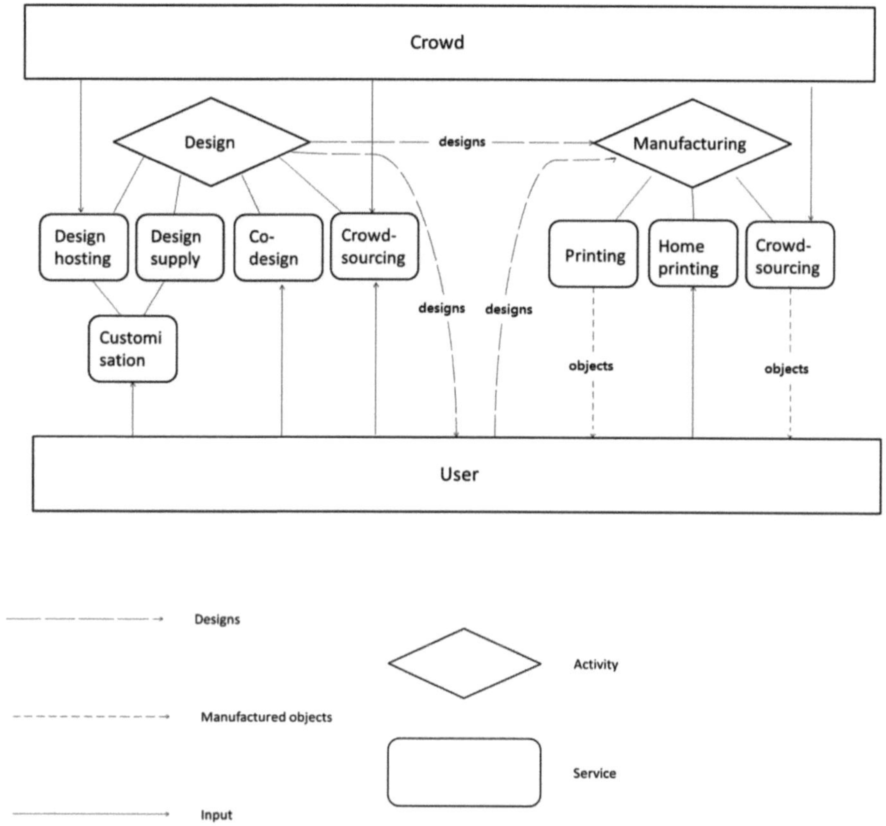

Fig. 7 Summary of design and manufacturing activities of online 3D platforms (Bogers et al. 2015)

4 Software for Additive Manufacturing

As there is a growing demand for AM, the number of software products available in the market has also risen substantially in recent decade.

4.1 Tinkercad

There is some software in the marketplace, which is specially designed for AM. Tinkercad is a user-friendly software that helps people design and make AM

Fig. 8 Tinkercad website (Tinkercad 2016)

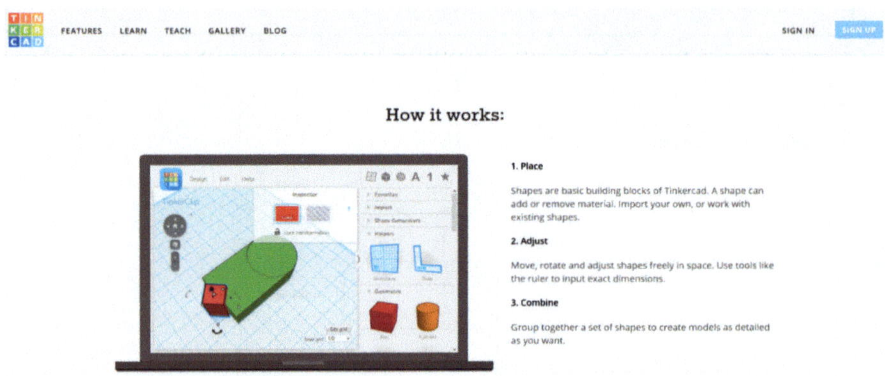

Fig. 9 Steps of using Tinkercad (2016)

product in a simple way with only three steps: placing, adjusting and combining. First, the user need to drag and drop different shapes from Tinkercad, or import shapes from other website or computers, in SVG format for 2D shapes or the STL format for 3D shapes. The second step is adjusting where users can move, rotate and adjust shapes freely. To measure the exact dimension, rulers are provided in this software. Figure 8 shows the screenshot of the homepage of Tinkercad, while Fig. 9 shows the screenshot of steps of using Tinkercad (2016). Figure 10 shows the application of Autodesk in creating 3D diagram for additive manufacturing.

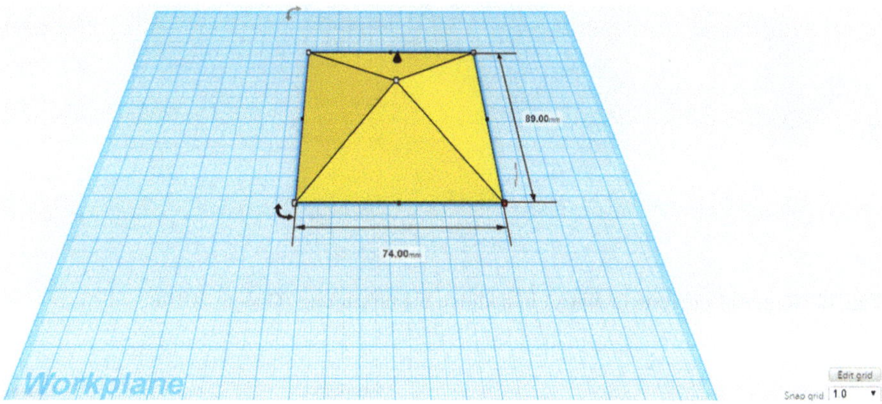

Fig. 10 Autodesk 123D (Autodesk 2017)

5 A Growing Trend in the Awareness of Additive Manufacturing: A Big Data Analysis

The data from Google shows that the trend of searches for AM from 2004 to 23 April 2016 rose dramatically (Fig. 11). In addition, the data shows that AM has already been focused on in many global areas. Figure 12 shows the regional interests (by country) in AM. The left side of the figure shows different countries and regions that are interested in AM and the numbers on the right side reflects the degree of interests for different regions. The higher the number, the greater is the degree of interest. Similar to Figs. 12 and 13 shows the regional interests (by city) in AM. The left side shows the cities interested in AM are mainly located in the southern hemisphere of the world, as indicated by blue circles in the map, while in the right side of the figure, the different numbers show their degree of interest in AM.

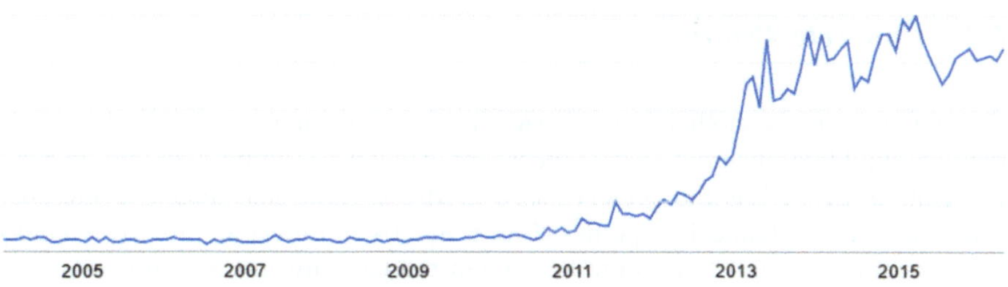

Fig. 11 Number of searches in Google from 2004 to 23 April 2016 (Google 2016)

Fig. 12 Regional interests (country) in additive manufacturing (Google 2016)

Fig. 13 Regional interests (cities) in additive manufacturing (Google 2016)

6 Methods Used for Additive Manufacturing in Construction Industry

Additive manufacturing has recently become accepted by the construction industry. There are three major printing forms used in the construction industry, AM contour crafting as previously discussed, concrete printing and D-shape.

6.1 Concrete Printing

Concrete printing is another form of AM. It uses extrusion technique to deposit the required building material. The process of concrete printing includes material preparation, as well as the cement-based mortar printing that meet specific characteristics and mechanical properties (Lim et al. 2012). Lim et al. (2012) concluded that concrete printing is based on the cement mortar extrusion, which is similar to contour crafting, but the process had been developed to retain three-dimensional freedom. This freedom is supported by the application of two build materials (one of the build material is used to create a structure similar to scaffolding that will be broken away after the completion of building process; a second material is deposited that is easy to remove).

6.2 *D-Shape*

Unlike the other methods that use cement-like paste to process AM, D-shape applies layers of powder or binder-jetting (Enshassi and Shakalaih 2015). This kind of AM also has a layer-upon–layer process. Each layer of building material is placed according to the chosen thickness, compacted. After that, the nozzles mounted on an AM gantry deposits the binder where it is to be solid. Once a proportion is completed, it is dug out from the loose powder bed (Lim et al. 2012).

6.3 *Comparison Between Contour Crafting, Concrete Printing and D-Shape*

The three major types of AM share some similarities. First, concrete printing and contour crafting are extrusion based. Concrete printing is realised with the help from cement-based paste, and contour crafting is made with cement mortar. Second, concrete printing needs to use the second material to support the created overhangs and other freeform features. The contour crafting and D-shape apply single material but use different methods to create different products (contour crafting does not need other materials to support created overhangs as it produces vertical elements largely in compression, while D-shape uses the unconsolidated material to support. Last but not least, these three kinds of AM share a common point, in that they are all layered process that build the product layer upon layer (Fig. 14) (Lee 2010).

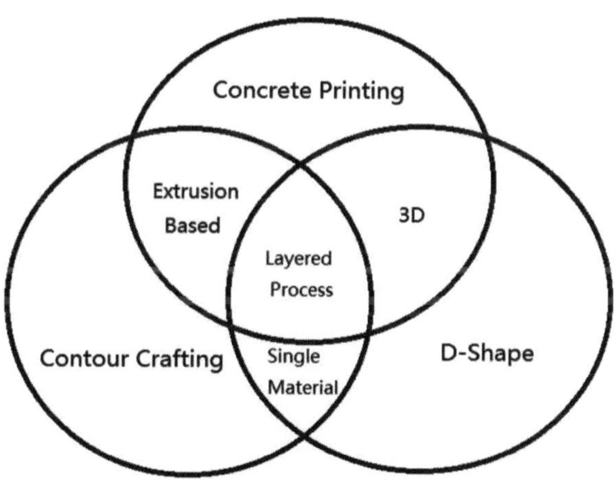

Fig. 14 Similarities between contour crafting, concrete printing, and D-shape (Lee 2010)

7 Some of the Sample Applications of Additive Manufacturing in Construction Industries

7.1 Three-Dimensional House Printing

Some countries have already applied AM to the building of houses. The pioneer process of AM in housing construction is called contour crafting, which is an addictive fabrication technology that uses concrete in the build (Weinstein and Nawara 2015). Lim et al. (2012) suggested that the principle of contour crafting was to use the cement-based past against a trowel to make multiple layers with smooth surfaces.

7.2 Three-Dimensional Bridges Printing

The technology of AM has also been applied in different countries for bridge construction. A bridge was built by AM technology. This bridge applied MX three-dimensional Resin technology, which was an enhanced technology that forms three-dimensional objects on any surface, independent of its inclination and smoothness, and in absence of any additional support structures (Duffy 2015). Marchand (2014) used AM material of White Strong Flexible (WSF) to make a model of the Palm River Bridge in Tampa, Florida, and the model bridge could not only be printed preciously, but also was strong enough to bare a normal human's weight. Nevertheless, Berman (2012) realised that in the US, bridge manufacturers applied AM when the tools were costly, complex and time consuming.

8 Costs and Benefits of Additive Manufacturing in Construction Industry

8.1 Benefits

Additive manufacturing offers a good way for RP as it is an automated process to produce models (Hague and Reeves 2000). Unlike the traditional modelling process, RP can save lots of time as it replaces the human modeller and toolmaker by using computer-based technology. Perkins and Skitmore (2015a, b) also indicated that AM could largely reduce the cost and improve building construction efficiency.

There are various advantages to incorporating AM concept in construction industry. First, integration of mechanical and electrical services within voids formed in the structure could optimise materials usage and site work. Second, there can be better control of the deposition of building material that produces novel internal and external finishes. Third, integrated units are created to reduce the

Table 2 Costs and benefits of additive manufacturing/3D printing

Author(s)	Benefit	Cost
Perkins and Skitmore (2015a, b)	/	The current 3D printing is unsuited to large-scale products and conventional design approaches and has a very limited range of materials that can be used
Bassoli et al. (2007)	It is effective in obtaining cast technological prototypes in a short time with It can be a low-cost construction approach with respect to previous technologies such as the selective laser sintering of sand Dimensional tolerances are completely consistent with metal casting processes Innovative solutions are available based on 3D printing process, which extend RC possibilities Layer-by-layer construction allows complex parts constructed, without any restrictions in terms of undercuts	/
Cesaretti et al. (2014)	3D printing can be used for direct construction of buildings and complex building structures with considerable dimensions	The cost of a spatial mission is very variable and depends on the destination, the level of safety as a consequence of the possible presence of men on board and obviously on the weight and volume of the material conveyed on the spatial vessel
Henke and Treml (2013)	It can be used in the fabrication of tailor-made building Envelopes or optimised building structures	/
Pfister et al. (2004)	Layer-by-layer construction of the scaffolds can be made by three-dimensional printing, i.e., bonding together starch particles followed by infiltration and partial crosslinking of starch with lysine ethyl ester diisocyanate	/

Table 3 Costs of three-dimensional printer

Manufacturer	Model	Features	Price
Three-dimensional systems	Cube 3	Build volume: 6 × 6 × 6 inches	$2327.42 + 405.2
		Materials: PLA and ABS	=$2732.62
		Resolution: 70 microns/layer	
		Heated bed: No	
		Multi-extrusion and auto-calibration	
		: Yes	
Model	Manufacturer	Features	$3685.01 + 809.63
Dremel	Three-dimensional idea builder	Build volume: 9 × 5.9 × 5.5 inches	=$4494.64
		Materials: PLA	
		Resolution: 100 microns/layer	
		Heated bed: No	
		Multi-extrusion: No	
		Auto-calibration: No	
Manufacturer	Model	Features	$1939.44 + 400.47
Printrbot	Simple	Build volume: 6 × 6 × 6 inches	=$2339.91
		Materials: PLA	
		Resolution: 100 microns/layer	
		Heated bed: Optional	
		Multi-extrusion: No	
		Auto-calibration: No	
Manufacturer	Model	Features	$1280.08 + 323.05
Tier time	UP mini	Build volume: 4.7 × 4.7 × 4.7 inches	=$1603.13
		Materials: PLA and ABS	
		Resolution: 200 microns/layer	
		Heated bed: Yes	
		Multi-extrusion: No	
		Auto-calibration: No	

(continued)

Table 3 (continued)

Manufacturer	Model	Features	Price
Manufacturer	Model	Features	$1280.08 + 323.05
Tier time	UP plus 2	Build volume: 5.5 × 5.3 × 5.5 inches	=$1603.13
		Materials: PLA and ABS	
		Resolution: 150 microns/layer	
		Heated bed: Optional	
		Multi-extrusion: No	
		Auto-calibration: Yes	
Manufacturer	Model	Features	$349.04 + 173.39
XYZ printing	da Vinci 1.0	Build volume: 7.8 × 7.8 × 7.8 inches	=$522.43
		Materials: PLA and ABS	
		Resolution: 100 microns/layer	
		Heated bed: Yes	
		Multi-extrusion: No	
		Auto-calibration: No	
Manufacturer	Model	Features	
XYZ printing	da Vinci 2.0	Build volume: 7.8 × 7.8 × 7.8 inches	
		Materials: PLA and ABS	
		Resolution: 100 microns/layer	
		Heated bed: Yes	
		Multi-extrusion: Yes	
		Auto-calibration: No	
Manufacturer	Model	Features	
XYZ printing	da Vinci 1.0 Jr. 1.0	Build volume: 5.9 × 5.9 × 5.9 inches	$4072.91 + 1745.57
		Materials: PLA	=$5818.48
		Resolution: 100 microns/layer	
		Heated bed: No	
		Multi-extrusion: No	
		Auto-calibration: No calibration needed	

(continued)

Table 3 (continued)

Manufacturer	Model	Features	Price
Manufacturer	Model	Features	
XYZ printing	da Vinci 1.1 Plus	Build volume: 7.8 × 7.8 × 7.8 inches	$230.8 + 157.88
		Materials: PLA and ABS	=$338.68
		Resolution: 100 microns/layer	
		Heated bed: Yes	
		Multi-extrusion: Yes	
		Auto-calibration: No	
Manufacturer	Model	Features	
XYZ printing	da Vinci 1.0 AiO	Build volume: 7.8 × 7.8 × 7.8 inches	$5422.89 + 629.65
		Materials: PLA and ABS	=$6052.54
		Resolution: 100 microns/layer	
		Heated bed: Yes	
		Multi-extrusion: Yes	
		Auto-calibration: No	

interface detailing and the likelihood of having expensive remedial works. Fourth, there can be a higher degree of freedom in designing in construction industry.

8.2 Costs

Additive manufacturing also has some disadvantages when applied within the construction industry (Buswell et al. 2007). Perkins and Skitmore (2015a, b) suggests that the current three-dimensional printing is unsuited to large-scale products and traditional design approaches, as is has a very limited range of materials which can be used. Other costs and benefits of 3D printing recognised in the literature are presented in the following Tables 2 and 3 identifies the current cost of 3D printers.

9 Research Method

In this research, in-depth interviews and a single case study was used. The in-depth semi-structured interviews results are read as follows:

9.1 Results of the Interviews

Most of the respondents in Hong Kong agree that AM is adopted in some construction projects. A safety officer from a gas company mentioned that his company is now exploring the possibility using AM to repair the existing gas conduit. They even have a lab for that but it is still at an experimental stage.

Nevertheless, many of the professors or construction personnel are quite conservative on making good use of 3D printing in the construction industry. A professor who has 30 years of experiences in construction automation suggested "*3D printing is a novel similar to what BIM was 30 years ago. Nobody is printing the whole building out in the US. I know that in China, there is a company claims to 3D print the whole house out. A [representative of a] housing company in Hong Kong wanted to visited them but they refused their entry. It is not what they are seen in the video. What I know is that, they can do non-structure. After traditional frame, they do the non-structure wall by 3D printing materials*".

He continued that "*In the United States, that's all at the experimental. For 3D printing, the main barrier is technical inability to put 3D printing to the contruction of load bearing components. Like you know you have to put several tons of weights to the materials, the 3D printing materials cannot withstand the load. In the laboratory, they have various different types of materials that they tried to tests the materials. They are leading to the goal that there is no problem in load bearing problems. In the United States, the 3D printing is not for practical use at all. There are no 3D printers which can print the whole house out. No one has tried to print the whole house out yet. In the East Coast, Clark construction is interested, they want to* [*start 3D printing for building structure*]".

A professor and department head from Colorado suggested, "*We are facing budget cuts in the United States. The financial status in the United States is not that good. Additive manufacturing is still in the infant stage. Most companies have not used additive manufacturing for their projects on sites. Most of the technology comes from the industry. We still have not used it to print the whole building out*". In Seattle, the CEO of a leading construction company suggests that they do not even consider to moving towards to 3D printing. They prefer to invest on construction robots and artificial intelligence.

9.2 Implications of Additive Manufacturing on Construction Safety

9.2.1 Additive Manufacturing Can Reduce Construction Accidents: Reduce Number of Manual Labours Work On-sites

A CEO of a small scale construction company in Hong Kong suggested that, "*additive manufacturing is a good idea, based on my experience on the construction work of CITY Bank. Even the three dimensional print success, there is still some human work needed. If workers follow the guidelines, the safety risk is low. However, if workers do not follow the guidelines, there is still an existing risk, even if we use the three dimensional print…Undoubtedly, if the three dimensional technology conducted by machine really works it can replace the human workforce. In addition, it can reduce the injury and cost. I think it is similar to the BIM*".

Similar to any other automated construction approach, interviewees suggested that AM's main benefit in construction safety lies on its ability to replace workers, a structural engineer in Hong Kong mentioned that, "*undoubtedly, if the three dimensional technology conducted by machine really works, it can replace the human workforce…it can reduce the injury and cost. I think it is similar to the BIM in reducing on site labour.*" A frontline worker in Hong Kong shares similar thoughts "*health and safety risks are reduced because many of the dangerous parts of the construction process will be eliminated*".

A researcher from Switzerland suggested that, "*3D printing…is more or less, in my view, somehow comparable to robotics. It can take out a lot of dangerous work that do not need to be done by workers. Everybody is excited about it. It provides a chance to increase the speed of production. Social housing and benefit will occur much quicker. Flying factory made parts from mainland China is cheaper that bringing raw materials from the factory, where the need to assemble the materials on site. They [the workers] said they are enthusiastic about this development, as this means they can complete the work in six months. It speeds up the processes and makes things more accurate*".

9.2.2 Additive Manufacturing Cannot Reduce Construction Accidents: It Is Only a Gimmick for Construction Companies

Some respondents took a dim view on the application of three-dimensional printing in the construction industry in Hong Kong. A safety officer considered that it is useless. Another safety officer in Hong Kong argued that AM could enhance safety on-sites theoretically speaking. Nevertheless, whether it can improve safety on-sites depend on frontline workers. If the tools are installed but the workers do not follow up well, safety risks still exist. That will be a problem when workers fail to follow company instructions.

A safety officer who has worked in the industry for 7 years, however, takes a different view on additive printing, "*I think additive manufacturing is not related to constructions but it is useful for building sales. It is not useful in relation to safety. I think this is hard to adopt in construction as it is hard to print the model and the workplace is keep changing*". A chairman of the labour union in Hong Kong also indicated that Hong Kong may not able to use AM, and further argued that it is a lot better to use precast construction, which can reduce accidents on-sites.

9.3 Costs of Additive Manufacturing

A construction company CEO in Fujian, China commented that, "*different materials and types cause different costs, my company has a 3D printer and it costs 1,200,000 RMB. We use it to do some exhibitions to our customers and it usually costs 5 or 6 h to print an object. The benefits usually above 1500 RMB.* Thus, costs can still outweigh the benefits. Nevertheless, it is used as a gimmick for exhibition".

A senior managerial member of a construction company in Seattle indicated that for their company, "*additive manufacturing can fasten the pace of construction processes, thereby saving money. Nevertheless, the barrier of adopting the technology is also costs. Construction practitioner can be motivated to use additive manufacturing if that can help us to reduce the construction processes*". When he was asked whether they have adopted AM, he admits that this is not the major priority in their company now.

An academic professor in Melbourne confirmed, "*The costs include the construction 3D printer and special concrete and materials for printing. It is cost effective and can create a new standard of housing construction in the future. It can reduce health and safety risks by replacing dangerous jobs on site with printing process*".

An innovation manager in one of the largest company suggests that the costs are quite low for printing the three-dimensional building out. They only use an ordinary three-dimensional printer, which can be bought in any relevant shop in Hong Kong. A frontline construction workers suggests that the, "*Cost is that any errors in the digital model could result in problems*". A CEO of a tool provider in Guangdong, however, considered that "*through the calculation and construct by the computer, the safety of the construction is ensured*".

9.4 Improvements in Communications Among Different Stakeholders

A BIM engineer in Hong Kong conceded that, "*we use 3D printing as that is easy to present to relevant stakeholders.*" Another practitioner from the same company does admit that, "*the usage of additive manufacturing is not only consider the construction safety, but also on site planning construction sequence, coordination and buildability. It helps them to rapidly understand the whole picture, rather than the traditional 2D drawing, despite the fact that this is time consuming*".

9.5 Better Quality Control

A project manager in Shanghai suggests that construction works can be more standardised by using AM, thereby increase the ease to conduct any analysis for the projects and give better advice An architect in Hong Kong agrees that "*Construction elements prefabricated or 3D printed in factories with quality control are of better quality and control than making structures on site. Labour hired on site could be reduced, and the chance of encountering construction accidents could also be reduced. The construction cost could sometimes be lowered because on site worker's salaries are, in general, expensive in Hong Kong. But it also depends on the cost of 3D printing and the fees for transportation*". A worker in Hong Kong also conceded that, "*construction will be faster and more accurate* [*with the help from additive manufacturing*]".

10 Case Studies Application in Construction Industry to Ensure Safety On-sites

In Hong Kong, most construction companies have not yet adopted AM. Two of the largest ones have provided a positive answer to this situation, simply suggesting that it is hard to make bricks without straws. These two companies admit that they have not yet used for printing out the whole building nor printing out small building parts. Yet, they do admit that they use the three-dimensional printer to print the original building model out to ease the discussion about the building process. The original motive, however, is nothing related to safety, as the three-dimensional printer is not very expensive. Nevertheless, it may take 24 h to print out one building structure (Figs. 15, 16, 17, 18, 19, 20, 21, 22, 23, 24 and 25).

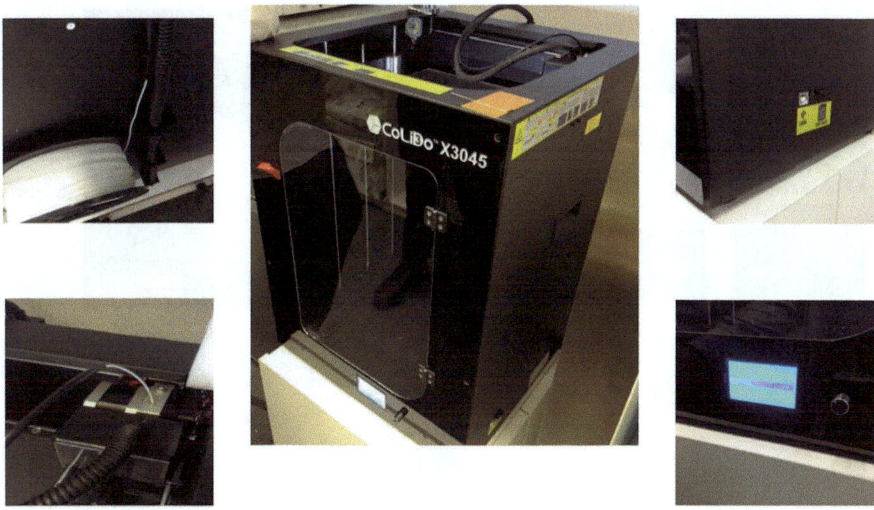

Fig. 15 In-house three-dimensional printer for printing building models (author's photos)

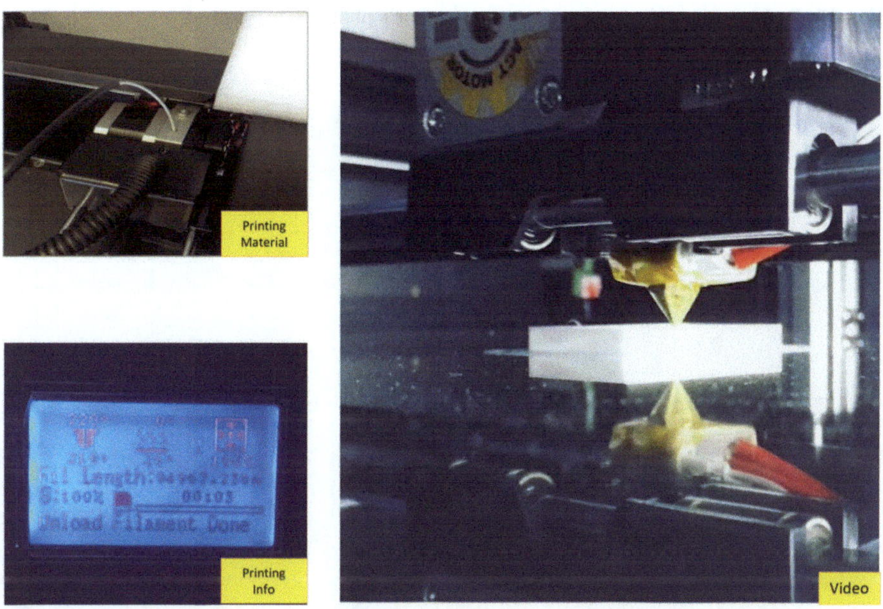

Fig. 16 Printing in process in three-dimensional printer (author's photo)

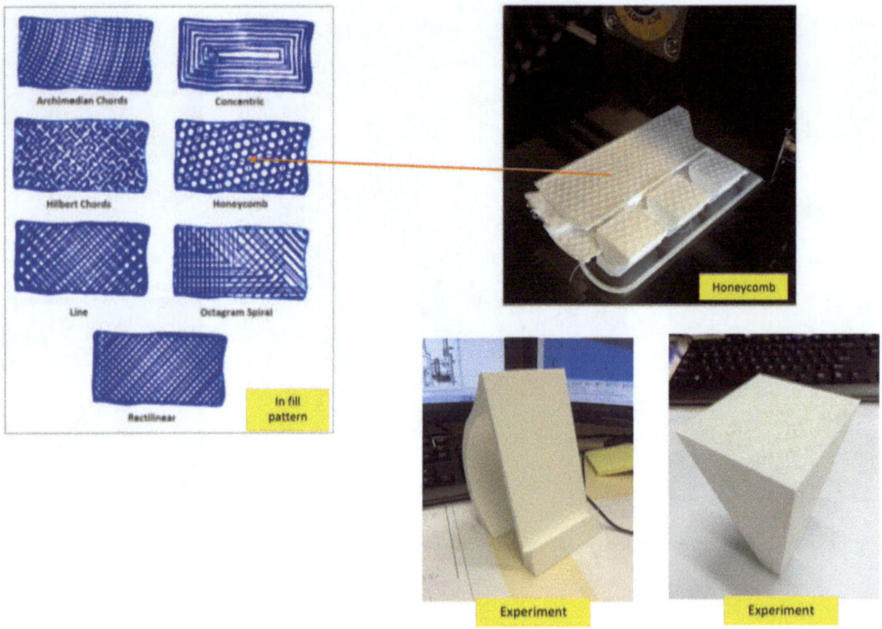

Fig. 17 Problems in additive manufacturing (author's photo)

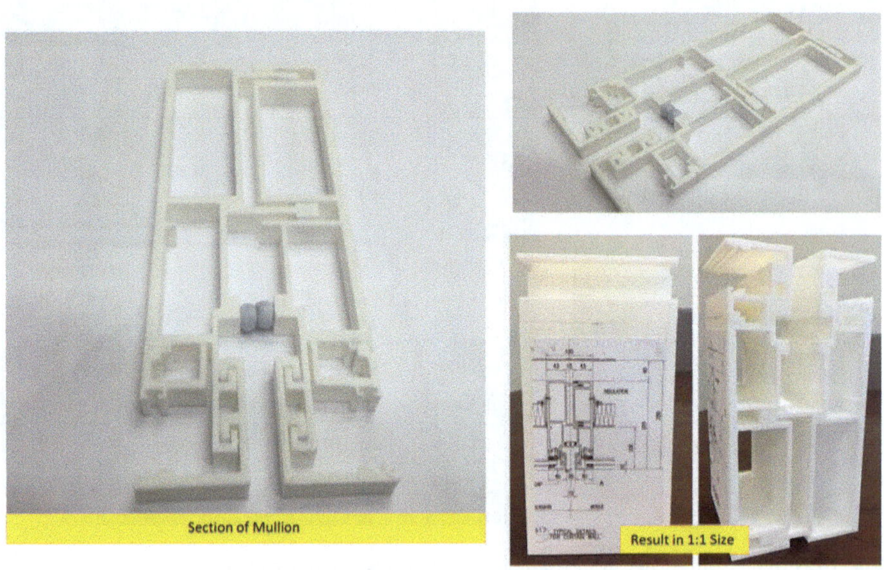

Fig. 18 Three-dimensional printed section of mullion (author's photo)

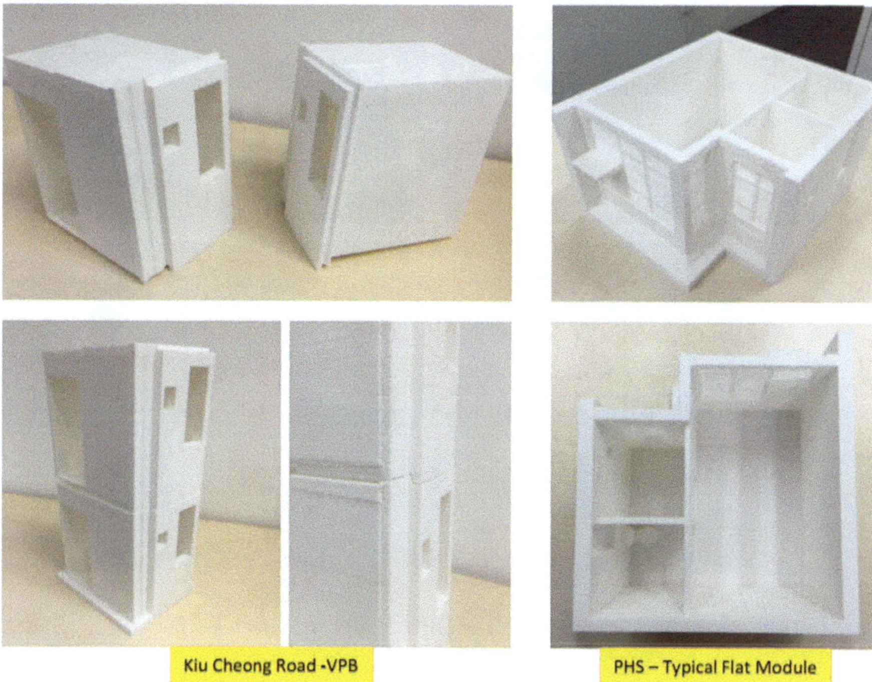

Fig. 19 Three-dimensional model of a typical housing unit (author's photo)

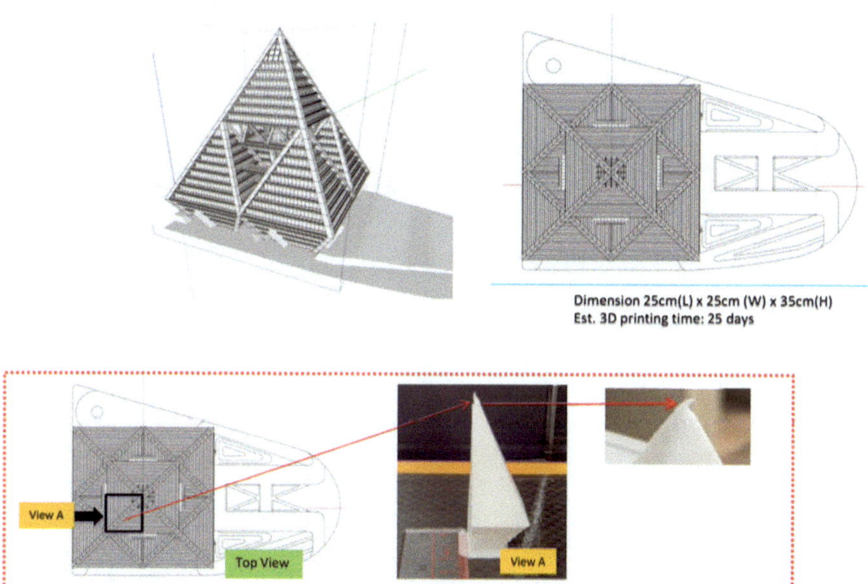

Fig. 20 Three-dimensional printed models of DPW Head Office (author's photo)

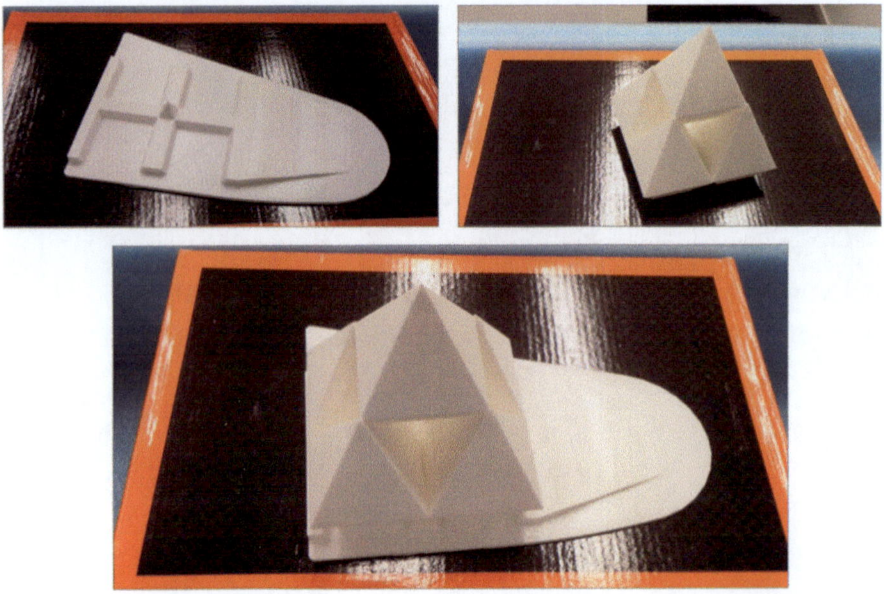

Fig. 21 Three-dimensional printed models of DPW Head Office (author's photo)

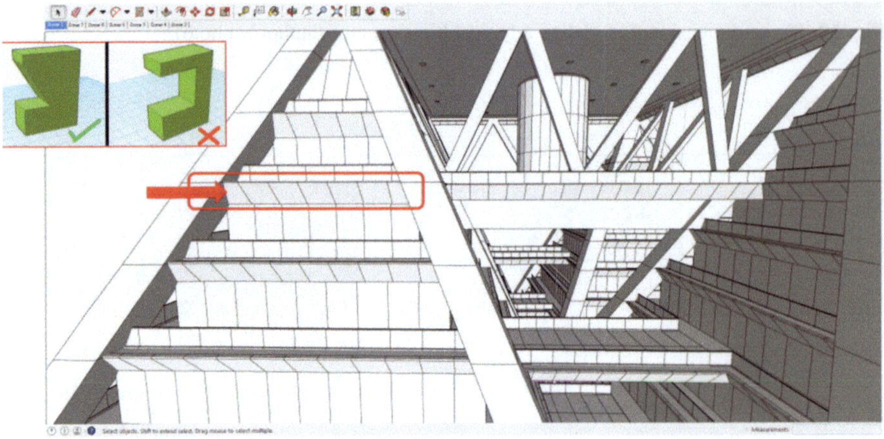

Fig. 22 Unable to print: hollow parts

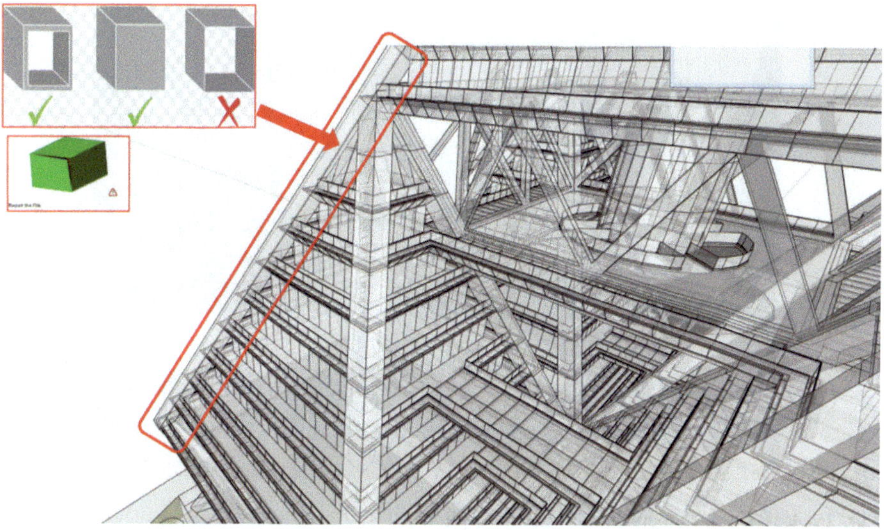

Fig. 23 Unable print overlap building structure (author's photo)

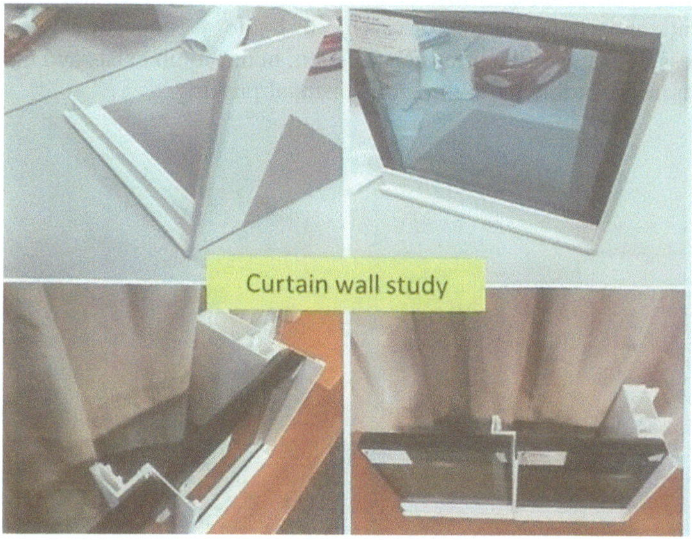

Fig. 24 Curtain wall additive manufacturing model (author's photo)

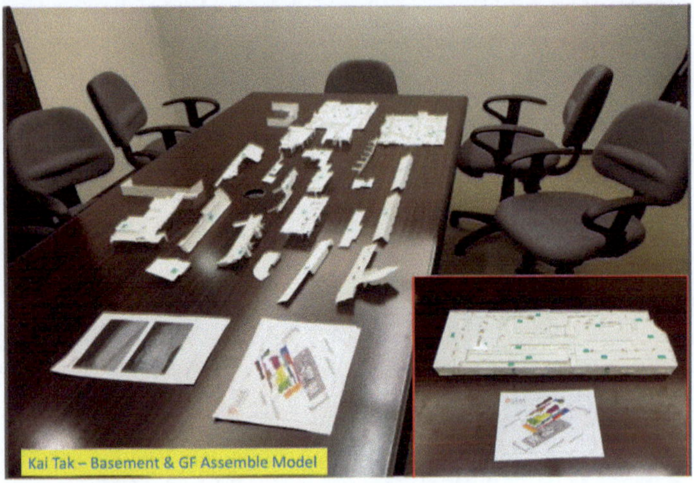

Fig. 25 Kai Tak basement model (author's photo)

The case study company admits that it has no special department for three-dimensional printing. Most of their colleague mainly comes from the BIM department and have an information technology educational background. Nonetheless, in some of the printing specimen, honeycomb exists and that becomes one of the major problems in three-dimensional printing.

Example of three-dimensional printed building structure Grand Lisboa

11 Conclusion

Recent technological breakthroughs have become an all-important economic issue across the whole range of industries, and the construction industry is no exception. In this context, AM portrays a new era in modular construction. Unprecedented complete 3D printed house buildings and civil engineering structures can be found around the globe. Lim et al. (2012) suggested that the driving factors that push forward the automation process (e.g., the development of AM and others automation technology) in the construction industry included a quest for improvement of construction safety and the reduction of construction time and costs. In particular, it is considered that with AM technology construction safety can be enhanced, which will also reduce both the direct and indirect causes of construction accidents.

With the technology of AM, the impact of weather on construction safety is lessened. The reason behind this is that the traditional manufacturing methods require a longer time of worker exposure on-site to finish the construction project, during which they can be subjected to inclement weather. Taking the building bridge as an AM example, onsite it requires time for the concrete to be hardened. When there is a sudden rainfall, apart from disturbing the effectiveness of the construction process, it is also likely to increase the levels of accidents occurring, such as those that might result from a landslide.

Acknowledgements The author would like to thank Tang, B's contribution in this chapter.

References

Autodesk. *Autocad Lt* 2017 [cited 24 April 2017]. Available from http://www.autodesk.com/store.

Bassoli, E., A. Gatto, L. Iuliano, and M. Grazia Violante. 2007. 3D printing technique applied to rapid casting. *Rapid Prototyping Journal* 13 (3): 148–155.

Berman, Barry. 2012. 3-D printing: The new industrial revolution. *Business Horizons* 55: 155–162.

Bogers, Marcel, Ian P. McCarthy, and Leyland Pitt. 2015. Leveraging users as innovators: Managing the creative potential of individual consumers. *Journal of Engineering and Technology Management* 37: 3–5. doi:10.1016/j.jengtecman.2015.09.001.

Buswell, R.A., R.C. Soar, A.G.F. Gibb, and A. Thorpe. 2007. Freeform construction: Mega-scale rapid manufacturing for construction. *Automation In Construction* 16 (2): 224–231.

Campbell, Thomas, Christopher Williams, Olga Ivanova, and Banning Garrett. 2011. Could 3D printing change the world?. *Atlantic Council Strategic Foresight Report*: 1–16.

Cesaretti, G., E. Dini, X. De Kestelier, V. Colla, and L. Pambaguian. 2014. Building components for an outpost on the Lunar soil by means of a novel 3D printing technology. *Acta Astronautica* 93: 430–450.

Chen, X., Q. Gao, X. Wang, and X. Li. 2016. Experimental design and parameter optimization for laser three-dimensional (3-D) printing. *Lasers In Engineering (Old City Publishing)* 33 (1–3): 189–196.

Duffy, Kaylie. 2015. 3D printing a functional steel bridge. *Engineering Answers*: 16–17.

Enshassi, Adnan, and Sobhi Shakalaih. 2015. Construction workers fall accidents from Scaffolding in Gaza Strip. *The 6th International Conference on Construction Engineering and Project Management (ICCEPM 2015)*.

Google. *Google Trend* 2016 [cited 2 August 2016. Available from https://www.google.com/trends/explore#q=3D%20printing.

Griffey, Jason. 2014. The types of 3-D printing. *Library Technology Reports*: 8–12.

Hague, R.J.M., and P.E. Reeves. 2000. Rapid prototyping, tooling and manufacturing. *Rapra Review Report* 10(9).

Henke, K., and S. Treml. 2013. Wood based bulk material in 3D printing processes for applications in construction. 71(1): 139–141.

HKSAR Labour Department. 2015. Occupational safety and health statistics bulletin. Issue No. 15. http://www.labour.gov.hk.

Hwang, Dooil, and Behrokh Khoshnevis. 2005. An innovative construction process-contour crafting (CC). *22nd International Symposium on Automation and Robotics in Construction ISARC 2005*: 1–6.

Jin, Yuan, Yong He, Fu Guoqiang, Aibing Zhang, and Du Jianke. 2017. A non-retraction path planning approach for extrusion-based additive manufacturing. *Robotics and Computer-Integrated Manufacturing* 48: 132–144. doi:10.1016/j.rcim.2017.03.008.

Kamali, Mohammad, and Kasun Hewage. 2017. Development of performance criteria for sustainability evaluation of modular versus conventional construction methods 142. *Journal of Cleaner Production* 142 (4): 3592–3606. doi:10.1016/j.jclepro.2016.10.108.

Khoshnevis, Behrokh. 2004. Automated construction by contour crafting—related robotics and information technologies. *Automation in Construction* 13 (1): 5–19.

Lee, An-Sheng. 2010. A study of the application of the Koch curve and Fuzzy theory on the computer aided design of furniture. *International Journal of Organizational Innovation* 3 (2): 438–465.

Li, Rita Yi Man, and Sun Wah Poon. 2013a. *Construction Safety*. Springer.

Li, Rita Yi Man, and Sun Wah Poon, eds. 2013b. *A literature review on the causes of construction accidents*. Germany: Springer.

Li, Rita Yi Man, and Sun Wah Poon. 2011. Using Web 2.0 to share the knowledge of construction safety as a public good in nature among researchers: The fable of economic animals. *Economic Affair* 31(1): 73–79.

Lim, Sungwoo, R.A. Buswell, T.T. Le, S.A. Austin, A.G.F. Gibb, and T. Thorpe. 2012. Developments in construction-scale additive manufacturing processes. *Automation In Construction* 21: 262–268.

Manocher, Djassemi. 2012. A computer-aided approach to material selection and environmental auditing. *Journal of Manufacturing Technology Management* 23 (6): 704–716. doi:10.1108/17410381211253290.

Marchand, Alex. 2014. Accessible 3-D printing. *Model Railroader* 81 (11): 63–65.

Matsunaga, Y.T., Y. Morimoto, and S. Takeuchi. 2011. Molding cell beads for rapid construction of macroscopic 3D tissue architecture. *Advanced materials* 23 (12): H90–H94.

Narayan, K. Lalit, K. Mallikarjuna Rao, and M.M.M. Sarcar. 2008. Computer aided design and manufacturing. *New Delhi: Prentice Hall of India*: 3.

Pan, W., and R. Sidwell. 2011. Demystifying the cost barriers to offsite construction in the UK. *Construction Management and Economics* 29: 1081–1099

Pei, Eujin, Jinsong Shen, and Jennifer Watling. 2015. Direct additive manufacturing of polymers onto textiles: Experimental studies and applications. *Rapid Prototyping Journal* 21 (5): 556–571.

Perkins, Isaac, and Martin Skitmore. 2015a. Three-dimensional printing in the construction industry: A review. *Journal: International Journal of Construction Management* 15(1): 1–9.

Perkins, Isaac, and Martin Skitmore. 2015b. Three-Dimensional printing in the construction industry: A review. *International Journal of Construction Management* 15(1): 1–9.

Pfister, A., R. Landers, A. Laib, U. Hübner, R. Schmelzeisen, and R. Mülhaupt. 2004. Biofunctional rapid prototyping for tissue-engineering applications: 3D bioplotting versus 3D printing. *Journal of Polymer Science Part A: Polymer Chemistry* 42 (3): 624–638.

Tinkercad. 2016. Tinkercad is a simple, online 3D design and 3D printing app for everyone https://www.tinkercad.com/.

Tuteski, Ognen, Atanas Kočov, and Taško Rizov. 2015. New product design development based on additive manufacturing & rapid prototyping methodology. *Journal For Technology of Plasticity* 40 (2): 55–64.

Weinstein, David, and Peter Nawara. 2015. Determining the applicability of 3D concrete construction (Contour Crafting) of low income houses in select countries. *Cornell Real Estate Review* 13: 94–111.

Wimmer, Rupert, Bernhard Steyrer, Josef Woess, Tim Koddenberg, and Norbert Mundigler. 2015. 3D printing and wood. *Pro Ligno* 11 (4): 144–149.

Chapter 5
Software Engineering and Reducing Construction Fatalities: An Example of the Use of Chatbot

Abstract The construction industry has historically done all it can to reduce the likelihood of accidents happening on sites. And, as it is virtually impossible to create one-size-fits-all solutions to such a complex problem area, based on one single piece of software, various types of software have been developed over the years for many different purposes relating to this. For example, some software systems have been designed to help safety officers investigate the safety conditions which pertain to a construction site. Other kinds of software facilitate quick drafting and designing and this can support safety on-site since it means that all changes can be recorded formally and centrally. Software engineering, therefore, has played an important role in fulfilling the needs of various kinds of construction practitioners. A case in point is that smart helmets, controlled by software, have recently been designed specifically to wake up workers who, the helmet has detected, have not been active for some time. Recent developments in software-driven user interfaces have meant that workers who have had a limited education can be informed of the facts they need to know on-site in a way that they can readily comprehend. Ease of use and the enhancement of safety have become two very important motivations for developers to create new tools and for the managers of construction sites to support their use, turning ideas which had previously been merely theoretical into things which are seen as achievable. In this paper, we present our first attempt at creating a simple chatbot application which can be used to share construction safety knowledge with the workers on a site. This is a pilot study. A simple conversation carried out via this software is given as an example of its use.

Keywords Software engineering · Knowledge-based systems · Fatalities · Construction industry · Chatbot

1 Construction Fatalities

When all goes well on a construction site, obviously fatal accidents do not happen. However, almost any system failure can lead to accidents and even fatalities. From 2013 to 2015, Colombia, Japan and Russia recorded more than 150 fatalities each,

R.Y.M. Li, *An Economic Analysis on Automated Construction Safety*,
DOI 10.1007/978-981-10-5771-7_5

annually. Other countries, such as Spain, accumulated more than 50 cases every year (Fig. 1). Eventually, contractors suffer economically from the very costly compensation claims which are made over the years (Li and Poon 2009a, b). For example, construction fatalities and injuries resulted in approximately US$15 billion in lost revenues every year in the United States alone (Tixier et al. 2016b).

Previous research shows that there is no consensus on what causes fatalities on sites. Although improper housekeeping is one of the major causes, we should never expect to see a construction site that is entirely neat and tidy. Next, workers can easily fall victim to accidents when they have not received sufficient training and have not had sufficient experience; this is especially the case in the construction industry which is notorious for its dangerous working conditions years (Li and Poon 2009a, b; Twentyman 2016). As it is quite often the case that the workers are paid via a piece rate, it is not unusual for workers to cut corners to save time. Also, loss of concentration or awareness is also culprits in terms of accidents—these problems are associated with over-long working hours Furthermore, wear and tear on facilities and equipment has always been a major cause of safety problems on construction sites (Twentyman 2016).

An important and under-researched cause of fatalities on construction sites is poor design. In Australia, it was found that poor design contributed to 44 percent of

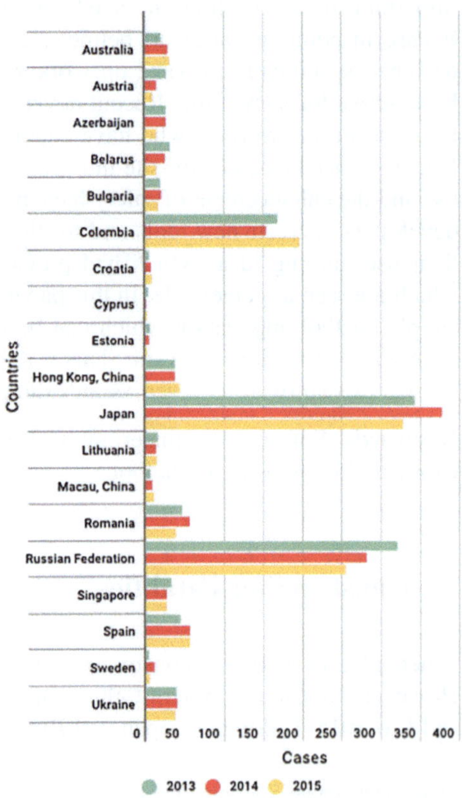

Fig. 1 Construction fatalities from 2013 to 2015 (International Labour Office 2017)

recorded work-related fatalities. The National OHS Strategy 2002–2012 states that the elimination of physical hazards at the design stage is a national priority. This strategy aims at enhancing safety awareness so that safety-aware design that can ensure safe outcomes can be recognised and aimed for. Accordingly, precise obligations for OHS designers of buildings structures have been included in the preventive OHS legislation in four Australian jurisdictions, i.e. South Australia, Western Australia, Victoria and Queensland. One way to improve design-related safety outcomes in the construction industry is for engineers and architects to conduct a thorough facilities' risk assessment (Cooke et al. 2008). Bong et al. (2015)'s desk study on pre- and post-harmonised WHS regulations reveals that designing for safety now constitutes a major theme in practice, following the harmonisation process. Designers are aware of the safety hazards on sites and design firms are willing to embrace the guidelines if they are protected from liability.

Hazard identification is the most frequently used approach to preventing and decreasing the number of fatalities on construction sites. Nevertheless, due to insufficient knowledge and/or experience of the exact circumstances which pertain, it is often difficult to identify hazards (Dzeng et al. 2016). Thus, many of the previous research studies throw light on the importance of construction safety knowledge sharing. Ling et al. (2009) suggest that effective communication between different stakeholders on construction sites is important for reducing fatalities.

2 The Role of Software Engineering in On-site Construction Safety

Eliminating the root causes that lead to construction fatalities is one of the major duties of all safety officers. They rack their brains to offer solutions to their contractors and developers which reduce the likelihood of accidents. In view of recent information technology developments, software engineering may be able to help (Abdulkhaleq et al. 2015). Having said that, as there are different safety requirements across the board due to the nature of construction sites, the weather and other site conditions, it is not easy to find a one-size-fits-all solution which is capable of identifying potential system hazards and enhancing safety management overall.

3 Software and Algorithms that Help Improve Construction Safety Performance on Sites

3.1 Geographical Information Systems (GIS)

Geographical Information Systems (GIS) enhance the in-house planning of construction projects. They reduce construction safety risks. They are capable of integrating diverse data sets, databases and applications together (Bansal 2011). GIS contributes to improvements in decision-making collectively among planners, designers and contractors. This technology masters both spatial and attribute information so that both can be quarried, analysed and displayed. Spatial data reveals feature geometries while attribute data, stored in tabular form, describes the characteristics of different features. GIS is about the various kinds of information which relate to building plans, site plans, drawings, subsurface detail plans, component specifications, building evacuation plans and landscaping plans. A GIS addresses construction safety since it maintains detailed information regarding the environment (Zhou et al. 2012).

3.2 Smart Helmets System

Smart helmets can be used to connect engineers in the field to more experienced colleagues at headquarters so that guidance can be provided through an audio and video link (Twentyman 2016). Using such bilateral communications, construction workers can obtain guidance which can alert them to dynamic changes in the working environment; thus, the risks of working on construction sites can be further minimised.

3.3 Virtual Reality and Augmented Reality

Virtual reality technology, alongside BIM technologies, can also be used to enhance construction workers' alertness in terms of construction safety. Virtual reality (VR) technologies have been used for training construction professionals on a risk-free but realistic, virtual, construction site. The Building Management Simulation Centre is a newly developed training centre for construction practitioners which allows them to have maximum control over the training processes and to obtain data about trainees which is gathered as they are training (De Vries et al. 2004; Zhou et al. 2012). With virtual reality technology, construction workers no longer need to access dangerous construction sites replete with a great many potential risks in order to train. They can now use VR technology at home and undertake their training in a convenient situation, and to repeat sessions as many times as they want in order to strengthen their skills. They can train for the new construction technologies that they will encounter in the foreseeable future.

A technology which is similar to VR, augmented reality (AR), plays an important role in enhancing construction workers' awareness at work. This technology presents a live view of a physical, real-world environment in which some elements have been augmented via computer-generated sensory input (Yoders 2014). AR has great potential to improve safety on construction sites when, for instance, operators are being trained on heavy equipment. With help from AR, workers do not need to receive training physically on-site. They can receive training first with augmented hazards produced by software and hardware. For example, workers can walk through an augmented construction site and learn to work with a full safety checklist (Cubedots 2016; Jones 2014) before they work on-site or receive physical training. This can avoid the problem of workers having insufficient training for them to be safe on-site, using the equipment that they need to. Different forms of training may also be used to support workers' memories in relation to the various relevant safety risks.

3.4 *On-site A.I Software for Use by Designers: Knowledge-Based Systems*

Knowledge-based systems (KBS) imitate human problem-solving via artificial intelligence techniques. They replicate human problem-solving expertise. KBSs are ideally suited to Occupational Health and Safety (OHS) decision support as this is a specialist area where learning about the mistakes which can be made is very desirable. The deployment of OHS expertise through software can be of significant benefit in OHS. Given that it is rare for experienced OHS staff to be in constant contact with construction design professionals such as architects and engineers, the provision of OHS decision support via a knowledge-based system has the potential to improve the designers' ability to integrate OHS into their design decision-making and OHS regulation compliance into the design stage (Cooke et al. 2008). Such a system provides designers and architects with expert knowledge on the safety performance of their designs by encoding within its expertise, standards and regulations that underpin given design problems (Cooke et al. 2008).

3.4.1 The Highway Concrete (HWYCON) Expert System

HWYCON is used by the highway departments to support their decisions with regard to the selection of materials, and to support their repair and recovery activities in regard to concrete structures (Cooke et al. 2008). The factors considered in the development of HWYCON included the following:

- Limit the scope of the knowledge domain, and then allow it to grow.
- Obtain feedback from users via prototype distribution and development.
- Selection of a development tool which provides a platform for further enrichments and inclusion of new knowledge.
- Selection of a development tool which allows flexibility for including various different forms of knowledge and ensures a high level of programming productivities (Kaetzel et al. 1994).

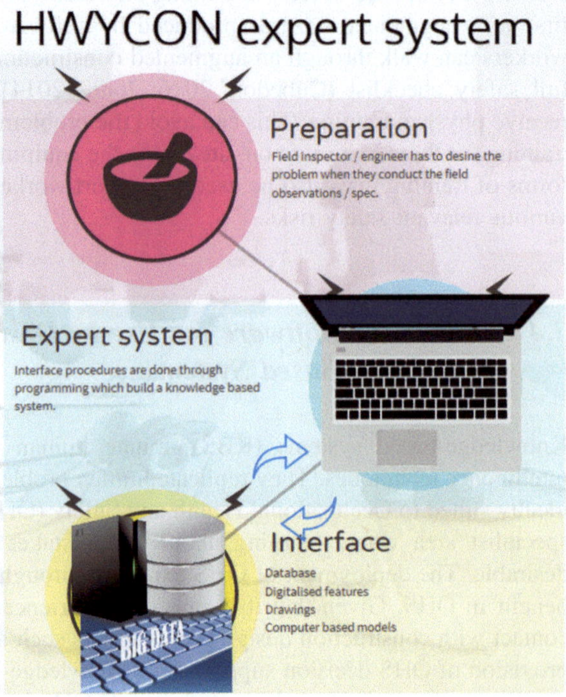

Fig. 2 Highway concrete expert system (Kaetzel et al. 1994) (author's figure)

The HWYCON system required field inspectors to study the problems, an expert system for which programming was needed in order to build the knowledge base, and an interface which allows access to a database, the CAD images and so on (Fig. 2).

3.4.2 Coronet

The Coronet system has been developed by Singapore's Building and Construction Authority for the purpose of applying artificial intelligence (AI) to the automating of the assessment, based on building regulations, of building plans. During automated plan checking, rules associated with each of the building entities are inspected to detect any breach of the building regulations. A prototype which deployed the Coronet technology was developed in order to deliver timely, knowledge-based advice on OHS in relation to building designs. The building's components are encoded and the OHS rules are applied to these in order to identify any risks which are associated with the design of each of the building entities (Cooke et al. 2008).

3.5 Computer Algorithms that Enhance Construction Safety

In addition, machine learning, random forest algorithms and stochastic gradient tree boosting are used to study the patterns that lead to construction accidents—in order to prevent construction injury (Tixier et al. 2016a). By using a hybrid simulation framework to combine discrete event simulation (DES) with system dynamics and agent-based simulation, a system has been implemented which reduces the dangers presented in the construction industry. This also speeds up construction time (Goh and Askar Ali 2016).

3.6 Can Software and Algorithms Enhance Safety Communications?

Safety communication refers to the information flow between safety, the environment and attributions regarding the causes of accidents (Motter and Santos 2017). Martins and Gorschek (2016) state that people mainly prefer an easy-to-use safety communication system. Systems include SafeML and SafeUML, can point out the communication among different stakeholders; however, the author also mentioned that the two new softwares need more research to test whether which one is better. In addition, Web 2.0 tools such as Wechat, Whatsapp and Facebook can also serve a similar purpose (Li and Poon 2011; Li 2015) and are now commonly used on sites.

4 Chatbot

Chatbots are machine conversation systems which interact with human users through natural, conversational language. Users interact with these applications by "chatting" with them. A variety of new chatbot technologies have arisen in recent years which simulate natural human language more accurately (Hill et al. 2015). Reshmi and Balakrishnan (2016) stated that "chatbot is a piece of software that responds to natural language input and attempts to hold a conversation in a way that imitates a real person". Some chatbots are used in commercial sector, whilst others are used for entertainment.

Turing proposed a behavioural approach to determining whether an interlocutor, via a terminal, was or was not a machine or artificial intelligence. An intelligent bot is considered to have passed the Turing Test when it is mistaken by a human assessor to be another human being in more than thirty percent of cases—after a five-minute question and answer session. In the so-called Standard Turning Test, a series of judges engage in chat interactions with two agents with the knowledge that one of them is human and the other is a "bot" using artificial intelligence. Turing maintained that the best test of machine sentience was to test whether a machine can fool human judges into believing that it was a human being (Gilbert and Forney 2015).

Chatbot technologies have different applications in different industries. Kerly et al. (2007) pointed out that chatbot technology could be applied to provide negotiation facilities and to enhance the open learner modelling environment. Chatbots may even have widespread applications in schools, universities and other training environments. Crutzen et al. (2011) realised that chatbots could be applied to answer adolescents' questions related to sex, drugs and alcohol. A chatbot could provide information in a positive and healthy way, especially in comparison to information lines and search engines. Shaw (2012) suggested that it would be easy to implement chatbots to answer the frequently asked questions (FAQs) currently listed on websites.

5 Costs and Benefits of Chatbot

Chatbots could be of benefit in a number of different ways. Jia (2009) indicated that if a chatbot were to be applied in language learning, it could make use of contextual information such as the dialogue context, the user's and its own personality knowledge, common sense knowledge, etc. This could provide tailor-made courses for each learner. Mou and Xu (2017) stated that people's personal traits tend to be expressed differently when they communicate with a chatbot instead of a human. People tend to be more self-disclosing, open, agreeable, extroverted and conscientious when interacting with humans rather than with an AI. Rossi and Carletti (2011) believed that chatbots could enhance teaching efficiency; they could be used to answer the most common questions which students ask as this relieves teachers and tutors from doing repetitive work. Chatbots, though, also have some disadvantages and costs. Hill et al. (2015) conducted a study on the differences between human–human online conversations and human–chatbot conversations. They found that chatbot communication lacks richness of vocabulary and may elicit greater profanity.

6 Making a Chatbot for Construction Safety Knowledge Sharing

We proposed that a chatbot could be made to share construction safety knowledge and replace human beings in this role. We further proposed that this could be achieved simply by inputting two sets of information to the Verbot 5 software, i.e. construction safety information for Hong Kong in English and construction safety information in Chinese. We then imitated a worker asking for information from our construction safety bot as follows Fig. 3.

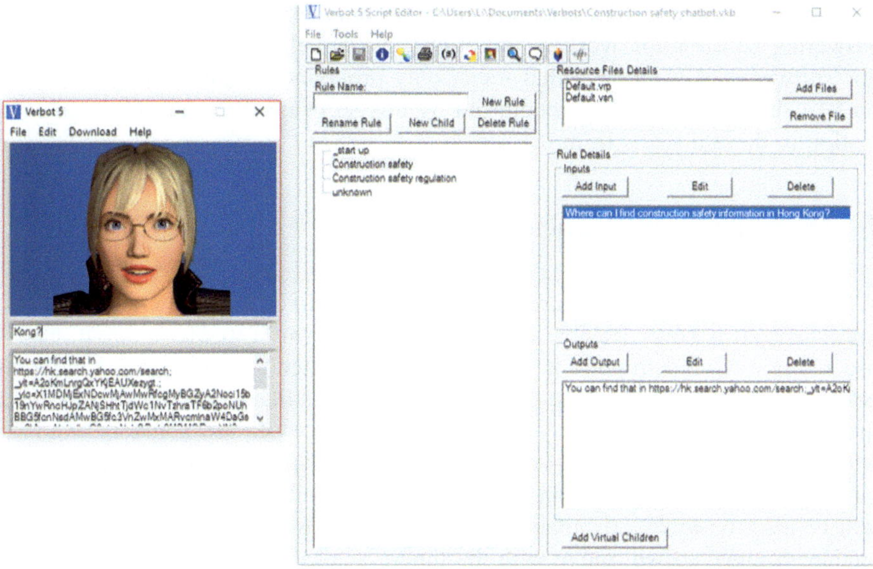

Fig. 3 Results when we ask exactly the same question "Where can I find construction safety information in Hong Kong?" (*Source* construction safety chatbot set up by the present author via Verbot 5)

Peter:	Where can I find construction safety information in Hong Kong?
Construction Safety Bot:	You can find that in https://hk.search.yahoo.com/search;_ylt=A2oKmLnrgQxYKjEAUXezygt.;_ylc=X1MDMjExNDcwMjAwMwRfcgMyBGZyA2Noci15b19nYwRncHJpZANjSHhtTjdWc1NvTzhraTF6b2poNUhBBG5fcnNsdAMwBG5fc3VnZwMxMARvcmlnaW4DaGsuc2VhcmNoLnlhaG9vLmNvbQRwb3MDMQRwcXN0cgNjb25zdHJ1Y3Rpb24lMjBzYWZldHklMjAEcHFzdHJsAzIwBHFzdHJsAzQwBHF1ZXJ5A2NvbnN0cnVjdGlvbiUyMHNhZmV0eSUyMGluJTIwSG9uZyUyMEtvbmcEdF9zdG1wAzE0NzcyMTQ3NDc-?p=construction+safety+in+Hong+Kong&fr2=sa-gp-hk.search&fr=chr-yo_gc&type=926458
Peter:	我們能在那裡找建築業安全的資料?(In Chinese, where can we find the construction safey information?)
Construction Safety Bot:	《建築地盤(安全)規例》可以在這裡http://www.legislation.gov.hk/blis_pdf.nsf/6799165D2FEE3FA94825755E0033E532/2CE1F50982C80299482575EE003561D7?OpenDocument&bt=0 找到

This demonstrates that chatbots may be able to replace human agents for answering construction workers' questions. Nevertheless, one major drawback to this software lies in the fact that it can only pronounce English words. The most popular dialects cannot be pronounced by this software. We can only read information in these dialects from the screen.

7 Discussion and Conclusion

When introducing new software, the user's perspectives must be considered. In a nutshell, as it is never feasible to shove software down a worker's throat, ease of use and minimum disruption in the workplace are often among the most important concerns on construction sites. Even though it sounds like a confidence trick to say that software is easy to use, since there is more easy-to-use software available, which feature facilities such as drag and drop, it is expected that the application of various types of software in construction processes will increase. More practitioners will be able to see the advantages brought by the advancements made in software engineering and so will move with the times.

In this present study, an attempt has been made to showcase the possible use of chatbot software for sharing knowledge via a computer system which replaces the functions of a human operator. Nevertheless, to make this exercise practical and meaningful, we may need to conduct extensive interviews with the construction safety officers and others involved so as to understand what their needs and requirements are in relation to the safety knowledge which may be asked for. In the near future, workers should be invited to validate the chatbot system and test whether it can effectively present safety knowledge to workers in the absence of any human operator or agent.

In addition, previous research shows that there are notable differences in the quality and content of conversations conducted with chatbots as opposed to with human beings. For example, people usually communicate with chatbots for longer, but with shorter individual messages than they do with other human beings. Also, human–chatbot communication usually exhibits less variety of vocabulary as compared to communications among people (Hill et al. 2015). All these issues indicate research questions which would be valuable to look into further.

Acknowledgements The author would like to thank Mr. H. Li, Mr. B. Tang and Mr. T. Mak's contribution in this chapter.

References

Abdulkhaleq, A., S. Wagner, and N. Leveson. 2015. A Comprehensive Safety Engineering Approach for Software-Intensive Systems Based on STPA. *Procedia Engineering* 128: 2–11.

Bansal, V.K. 2011. *Application of geographic information systems in construction safety planning*. 29 (1): 66–77.

Bong, S., R. Rameezdeen, J. Zuo, R.Y.M. Li, and G. Ye. 2015. The designer's role in workplace health and safety in Construction Industry: post-harmonized regulations in South Australia. *International Journal of Construction Management* 15 (4): 276–287.

Cooke, T., H. Lingard, N. Blismas, and A. Stranieri. 2008. ToolSHeDTM: The Development and Evaluation of a Decision Support for Health and Safety in Construction Design. *Engineering, Construction and Architectural Management* 15 (4): 336–351.

Crutzen, R., G.-J.Y. Peters, S.D. Portugal, E.M. Fisser, and J.J. Grolleman. 2011. An Artificially Intelligent Chat Agent That Answers Adolescents' Questions Related to Sex, Drugs, and Alcohol: An Exploratory Study. *Journal of Adolescent Health* 48 (5): 514–519.

Cubedots. 2016. How augmented reality in real estate will benefit the construction industry.

De Vries, B., S. Verhagen, and A.J. Jessurun. 2004. Building Management Simulation Centre. *Automation in Construction* 13 (5): 679–687.

Dzeng, R.-J., C.-T. Lin, and Y.-C. Fang. 2016. Using eye-tracker to compare search patterns between experienced and novice workers for site hazard identification. *Safety Science* 82: 56–67.

Gilbert, R.L., and A. Forney. 2015. Can avatars pass the Turing test? Intelligent agent perception in a 3D virtual environment. *International Journal of Human-Computer Studies* 73: 30–36.

Goh, Y.M., and M.J. Askar Ali. 2016. A hybrid simulation approach for integrating safety behavior into construction planning: An earthmoving case study. *Accident Analysis and Prevention* 93: 310–318.

Hill, J., W. Randolph Ford, and I.G. Farreras. 2015. Real conversations with artificial intelligence: A comparison between human–human online conversations and human–chatbot conversations. *Computers in Human Behavior* 49: 245–250.

International Labour Office. 2017. Construction Accidents.

Jia, J. 2009. CSIEC: A computer assisted English learning chatbot based on textual knowledge and reasoning. *Knowledge-Based Systems* 22 (4): 249–255.

Jones, K. 2014. Five ways AR will benefit the construction industry. ***News***.

Kaetzel, L. J., J.R Clifton, and K. Snyder. 1994. Users Guide to the Highway Concrete (HWYCON) Expert System.

Kerly, A., P. Hall, and S. Bull. 2007. Bringing chatbots into education: Towards natural language negotiation of open learner models. *Knowledge-Based Systems* 20 (2): 177–185.

Li, R.Y.M. 2015. Construction Safety Knowledge Sharing via Smart Phone Apps and Technologies. In *Handbook of Mobile Teaching and Learning*, ed. Y. Zhang, 1–11. Berlin, Heidelberg: Springer, Berlin Heidelberg.

Li, R. Y. M., and S.W. Poon. 2009a. Future Motivation in Construction Safety Knowledge Sharing by Means of Information Technology in Hong Kong. *Journal of Applied Economic Sciences* 4 (3(9)): 457–472.

Li, R.Y.M., and S.W. Poon. 2009b. Workers' Compensation for Non-fatal Construction Accidents: Review of Hong Kong Court Cases. *Asian Social Science* 5 (11): 15–24.

Li, R. Y. M., S. W. Poon. 2011. Using Web 2.0 to Share the Knowledge of Construction Safety as a Public Good in Nature Among Researchers: the Fable of Economic Animals. *Economic Affair* 31(1): 73–79.

Ling, F.Y.Y., M. Liu, and Y.C. Woo. 2009. Construction fatalities in Singapore. *International Journal of Project Management* 27 (7): 717–726.

Martins, L.E.G., and T. Gorschek. 2016. Requirements engineering for safety-critical systems: A systematic literature review. *Information and Software Technology* 75: 71–89.

Motter, A.A., and M. Santos. 2017. The importance of communication for the maintenance of health and safety in work operations in ports. *Safety Science* 96: 117–120.

Mou, Y., and K. Xu. 2017. The media inequality: Comparing the initial human-human and human-AI social interactions. *Computers in Human Behavior* 72: 432–440.

Reshmi, S., and K. Balakrishnan. 2016. Implementation of an inquisitive chatbot for database supported knowledge bases. *Sadhana* 41 (10): 1173–1178.

Rossi, P.G., and S. Carletti. 2011. MAPIT: A pedagogical-relational ITS. *Procedia Computer Science* 3: 820–826.

Shaw, A. 2012. Using Chatbots to Easily Create Interactive and Intelligent FAQ Webpages. *Journal of Applied Global Research* 5 (15): 10–15.

Tixier, A.J.P., M.R. Hallowell, B. Rajagopalan, and D. Bowman. 2016a. Application of machine learning to construction injury prediction. *Automation in Construction* 69: 102–114.

Tixier, A.J.P., M.R. Hallowell, B. Rajagopalan, and D. Bowman. 2016b. Automated content analysis for construction safety: A natural language processing system to extract precursors and outcomes from unstructured injury reports. *Automation in Construction* 62: 45–56.

Twentyman, J. 2016. Wearable devices aim to reduce workplace accidents: *Financial Times*.

Yoders, J. 2014. How is augmented reality being used in construction, *Building Design*, Vol. 2016: Line//Shape//Space.

Zhou, W., J. Whyte, and R. Sacks. 2012. Construction safety and digital design: A review. *Automation in Construction* 22: 102–111.

Chapter 6
Virtual Reality and Construction Safety

Abstract Virtual reality (VR) simulates real-world situations. Players are immersed in an environment that resembles reality. In the past, head-mounted VR devices' costs were insurmountably high, nipping their application in the bud. Thus, VR is not common among general laymen. In recent years, however, technological breakthroughs in smartphones and VR gear have turned a new page in the gaming industry. Handy software and hardware have turned VR into an affordable gaiety of nations. It has been included in the electronic gaming industry in recent years. It has also been applied in car-driving lessons and shopping mall promotional activities. VR, which is adopted in the construction industry for safety training, is also as fine as a ninepence. In this chapter, we aim to study (1) the popularity of searches of VR around the globe, (2) the costs and benefits of VR in the construction industry and (3) real-life applications of the tools in Hong Kong for safety training and the US construction industry when the building is still in its inception stage.

Keywords Virtual reality · Construction safety · Safety training

1 Introduction

The construction industry creates lots of job opportunities and contributes much to our economy every year. It is usually one of the major powerhouse in economic growth (Li et al. 2015). Although the economic benefits are clear with a multitude of job opportunities, the construction industry, in general, has a poor safety record around the globe and the dangers involved are notorious (Grant and Hinze 2014; Li et al. 2017, forthcoming). Construction injuries often ruin the lives of victims and leave everyone around them worried. How to reduce accidents on-site has always been considered a sticky issue.

In the United States, incident rates in the construction industry are twice the industrial average. The United States National Safety Council reported that there are nearly 2200 deaths and 220,000 disabling injuries each year. In Japan, construction fatalities account for 30–40% fatal industrial accidents, and half in Ireland. In the

R.Y.M. Li, *An Economic Analysis on Automated Construction Safety*,
DOI 10.1007/978-981-10-5771-7_6

United Kingdom, major injuries reported amounted to more than 3000 in 2005/2006 in the construction sector. In Europe, even though only 10% of the population is employed in the construction industry, 30% of fatal industrial accidents are related to the construction industry (Li et al. 2017, forthcoming).

In some developing nations, construction in developing countries, especially in South Asian ones such as India, Pakistan and Sri Lanka, is labour intensive. Most workers are unskilled and migrants. Communication problems due to language and cultural differences lead to compromised safety on-site (Azhar and Choudhry 2016). For example, in Saudi Arabia, more than half of the total injuries at work are related to the construction industry (Li et al. 2017, forthcoming). In Pakistan, Bailenson et al. (2008), injuries in the construction industry were the second highest among all industries (Azhar and Choudhry 2016).

High accident rates often lead to a delay or extension of time in the construction industry due to accidents causing investigations. Costs also rise due to lengthening construction time. Thus, a vast amount of research has been conducted to research the causes of accidents. Previous research has indicated that construction accidents are affected by the site layout, materials, tools and equipment, management and workers on-site (Li and Poon 2013a; Lingard et al. 2017). Some researchers have found that better design (Bong et al. 2015; Grant and Hinze 2014), planning, safety management and training enhance safety on-site and prevent construction accidents. They help site managers and workforces to identify and recognise safety risks and enhance communication (Li and Poon 2013a; Li 2015b; Park and Kim 2013).

Literature argues that effective training programs will directly affect human error. Poor training contributes to more than half of occupational incidents. Thus, it is the nuts and bolts of construction safety. Previous research has found that there is a gap between the expectation of the construction industry and the status of the conception of safety. To offer a safe working environment, the VR-based safety training program is a platform on which users can rehearse tasks and promote their abilities through electrical hazards. Its simulation and visualisation features remove the training barriers due to potential safety hazard and invisibility (Zhao and Lucas 2015).

2 Virtual Reality

In 1957, Morton Heilig invented the Sensorama, a simulator with three-dimensional images to simulate real-life situations. That was the first generation of VR. In 1965, Ivan Sutherland used a head-mounted display (HMD) connected to a computer, allowing users to see a virtual world. Because of the high costs, head-mounted devices (HMD) for virtual reality (VR) were only adopted by large companies, and fewer users could gain experience in these systems (Hilfert and König 2016). Nevertheless, technological breakthroughs have meant that VR gear can be viewed from smartphones. It has become cheap enough to be a toy for everybody. Construction workers are no exception. Details of VR's history is shown in Fig. 1.

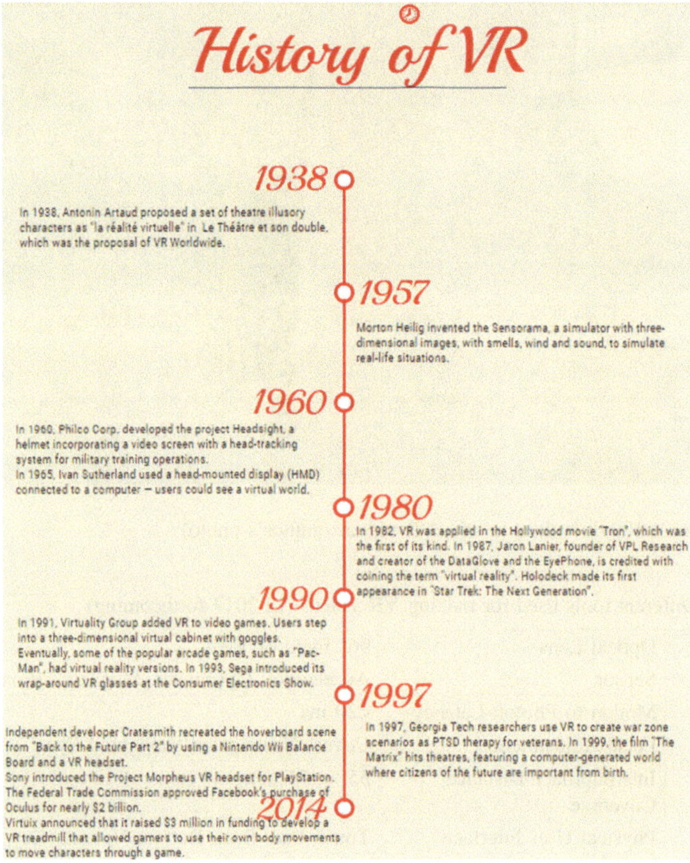

Fig. 1 History of VR (Artaud 1938)

Virtual reality (VR) needs a computer, software and hardware to build a simulated environment for its user. It can be a realistic or imaginary environment. Realistic images can be taken by a VR camera or even a robot, such as Double DJI, 360 Panorama and so on (Li 2017, forthcoming-a (Fig. 2 and Table 1)). An immersive virtual environment (IVE) is a computer-generated environment that can engage the person's senses and remove their perception. An IVE creates the sense of "being there" in a three-dimensional virtual environment. It can track the user's location and orientation to update the virtual scene, which matches the user's movements; it gives the user some degree of control over the objects therein (Bailenson et al. 2008). Head and body movements are a vital input of a virtual avatar. The techniques are similar to those of augmented reality (AR). VR is able to present models and scenes using 3D techniques that attract a wide range of audiences (Hilfert and König 2016).

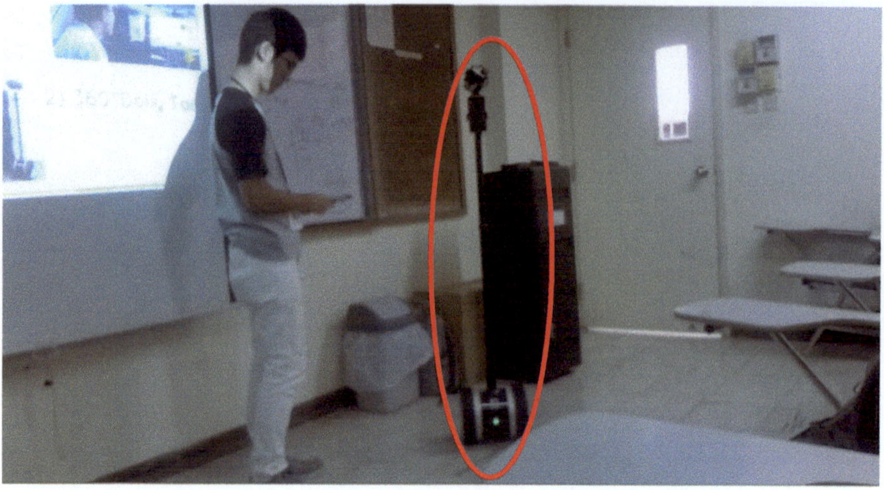

Fig. 2 Double Robot is used to take VR videos (author's photo)

Table 1 Different tools used for making VR videos (Li 2017 forthcoming)

Gear VR	Optical Lens	96° Field of View
	Sensor	Accelerator, Gyrometer, Magnetic, Proximity
	Motion to Photon Latency	<20 ms
	Focal Adjustment	Covers Nearsighted/Farsighted Eyes
	Interpupillary Distance Coverage	55 ~ 71 mm
	Physical User Interface	Touch Pad, Back Button, Volume Key
	Connection	Micro USB 1.1 Connection to the Galaxy Note 4
	Dimension (Headset)	198 (W) × 116 (L) × 90 (H) mm

A central notion of VR is the feeling of being there in the virtual environment, i.e. subjects' responses in the virtual environment are similar to their responses in a real environment. VR, which is used to train workers, has become common in recent years. For example, flight simulators are using VR to show the visual environment for road-safety training. The reason for VR becoming common in training workers is that VR can avoid the dangers inherent in the training of humans or animals (Sacks et al. 2013).

3 Popularity of VR as Reflected in the Number of Google Searches: Big Data Analysis

We adopt big data analytics, which can act as a proxy in studying the popularity of virtual reality in different parts of the world from 2004 to 2016. It is the use of data collection and analysis which optimise decision-making. It reviews a large volume

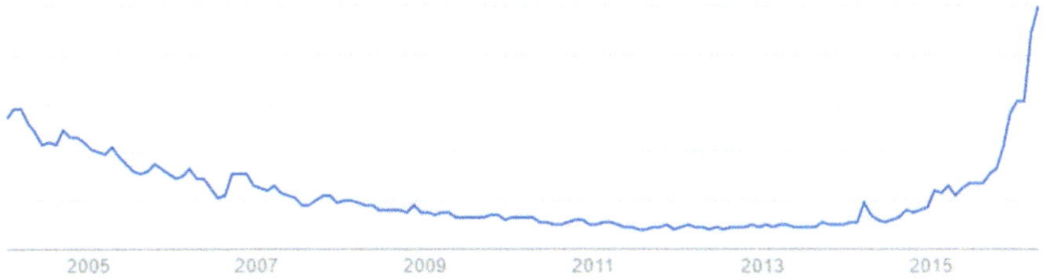

Fig. 3 Number of "virtual reality" searches in Google (Google 2016)

Fig. 4 Countries with the largest number of searches around the globe (Google 2016)

of unorganised data to identify patterns, helping decision makers or organisations to attain a better understanding of behaviours (Li et al. 2016). In this chapter, we study the popularity of VR as reflected in the number of searches in Google, as shown in Fig. 2. The largest number of searches came from Singapore, Malaysia and the United States (Fig. 3) and Melbourne, Sydney and Chennai (Fig. 4). Most of the related searches are related to the tools of virtual reality such as Oculus (Google 2016).

4 VR Applications

4.1 Gaming Industry

In recent years, virtual reality in computer games and mobile phones has been gaining popularity. The year 2016 has been coined the VR year due to the wide commercialisation of VR technology in the film, entertainment, video game and news industries. VR and video games have boomed in the entertainment industry. The interactivity in video games, allowing real-time feedback in simulated, real-life situations, is collected from a communications channel's receivers. It continually modifies the message being delivered to the receiver (Lin 2017).

The challenges shown in a virtual reality game enhance people's learning on top of the attraction of extra engagement. Learners may proceed from a lower level to a

Fig. 5 Cities with the largest number of searches around the globe (Google 2016)

higher level according to their capability (Hamari et al. 2016). The hypothesising, probing and reflecting process in the simulated world enhances learners' motivation (Hamari et al. 2016).

4.2 Driving Simulations

Most VR studies recorded previously only focus on the vehicle-driving simulation system. Recently, many users have adopted simulation technologies for learning. Drivers can improve their driving skills and know more about their problems in

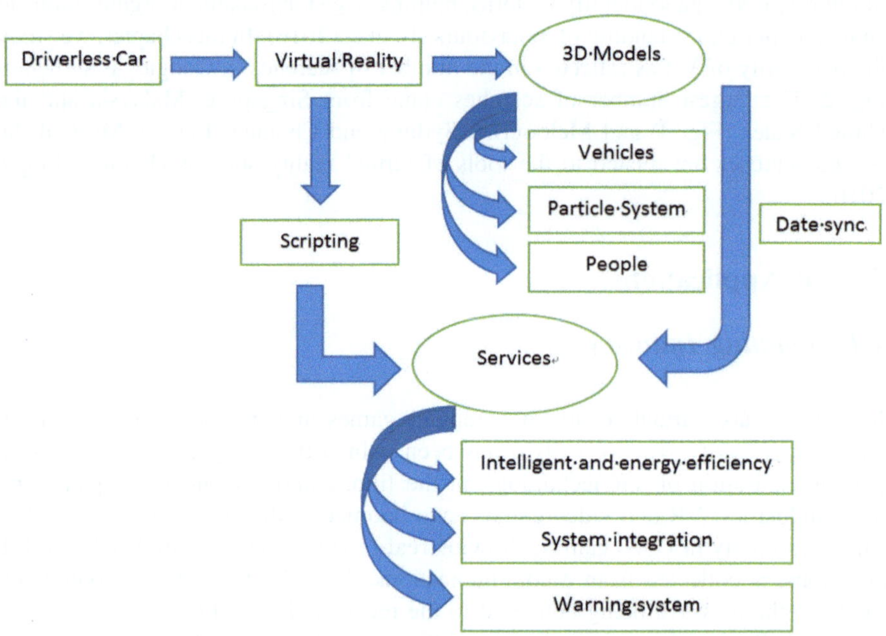

Fig. 6 Typical VR safety training simulation (Hsu 2016)

Fig. 7 Tennis simulation game in shopping mall (author's photo)

driving. This approach reduces the likelihood of repeated driving problems (Hsu 2016). The idea of VR car simulations can be found in the following Fig. 5.

4.3 Shopping Mall Promotions

Apart from driving simulations, VR is also used in a number of shopping malls in Hong Kong and Mainland China to increase the number of shoppers and boost business profits of shops. For example, one of the shopping malls has introduced VR tennis for shoppers who have spent over a certain amount of money in the mall, as shown in the following Figs. 6 and 7.

4.4 VR Application in Teaching and Learning: An Example of Edutainment

Apart from traditional classroom learning settings, learners' engagement can be enhanced by VR environments (Hilfert and König 2016). In education, the idea of 'edutainment', accessing learning content via both an educational and enjoyable environment, enhances students' learning interests. For example, VR provides students with valuable experience personally in old, urban areas without an actual visit in person.

Previous research has stressed the importance of learning in different environments. Thus, schools organise different field trips, e.g. farms, Wetland Park or even places out of town, to allow students to immerse in different learning environments. Having said that, however, many of the outing activities may cost learners an arm and a leg, posing an insurmountable financial burden on parents or even schools. In view of the resource limitation, such as manpower, budget and time, schools cannot arrange as many field trips for every module.

With the help of VR gear and smartphone devices, students can visit various places in a flash. They can explore the environment independently in a lively and interesting manner. Students can compare and contrast the similarities and differences between two places instantly without travelling. Teachers can hold a discussion immediately after the VR tour. Some of the teachers reported that learning via VR is far more efficient and effective than traditional teaching (Li 2016).

4.5 VR Application in Construction Industry

According to the nature of construction sites and previous experience of failure, purpose-built training facilities that physically simulate the construction site (de Vries et al. 2004) are often built. They are available in the United Kingdom, Australia, the Netherlands and Hong Kong (Li and Poon 2013b). In VR training, workers can decide on a course of action, implement the action and immediately observe the results. This results in cognitive information processing, which leaves its mark in long-term memory (Lucas et al. 2008). Lucas et al.'s (2008) research is focused on the safety of training for operators of construction equipment.

Planning before working on-site helps to identify the risk factors and safety information according to project activities on-site. It can be performed via meeting with safety managers and various stakeholders using pre-designed virtual project sites in the graphic library. Safety risks, unsafe conditions, accident cases and checklists can be extracted

Fig. 8 VR game store for shoppers in a shop in Tseung Kwan O, Hong Kong (author's photo)

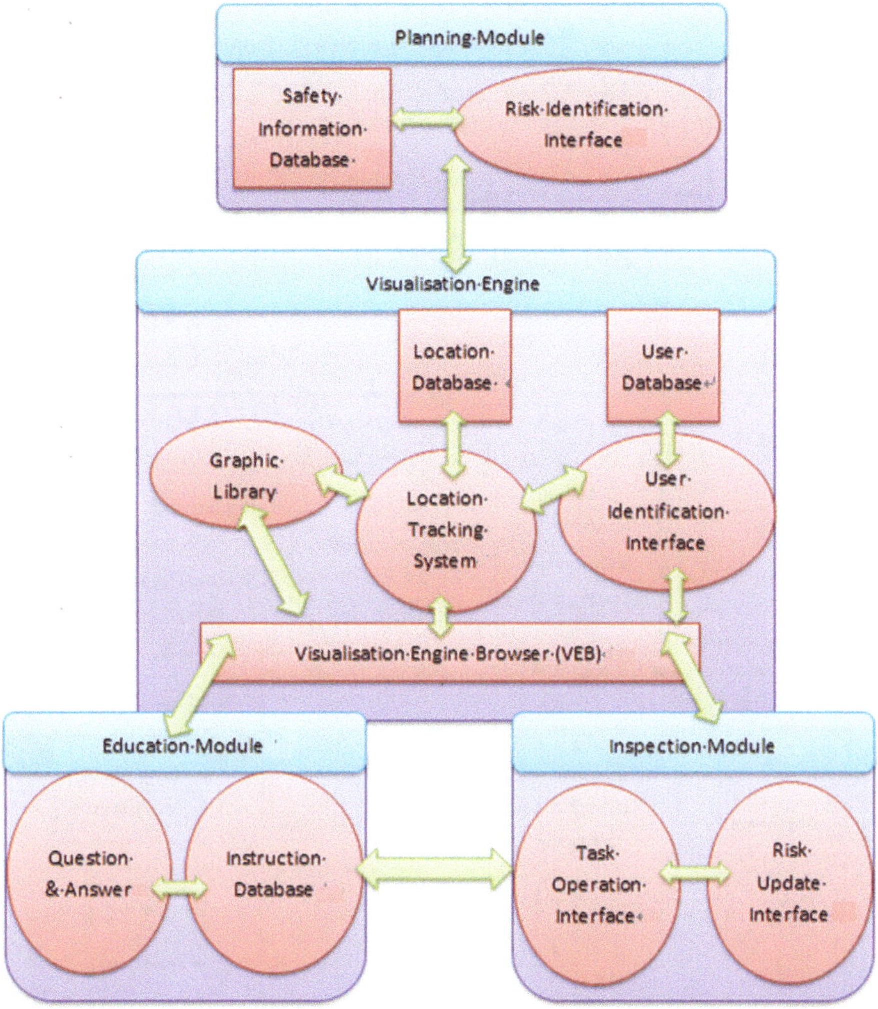

Fig. 9 Example of VR application with planning, visualization, education, inspection modules in constructionindustry

from a safety information database. The information can then be used for making virtual site models. Whenever site inspectors and officers complete some of the education and inspection tasks, the database will be updated (Fig. 9). A risk identification interface (RII) allows a safety manager to know more about workers' safety education and the planned field inspection. New risk factors that are found during the safety inspection process can be stored automatically in the database (Park and Kim 2013) (Fig. 10). The figure below illustrates the applications of VR in different industries (Fig. 11).

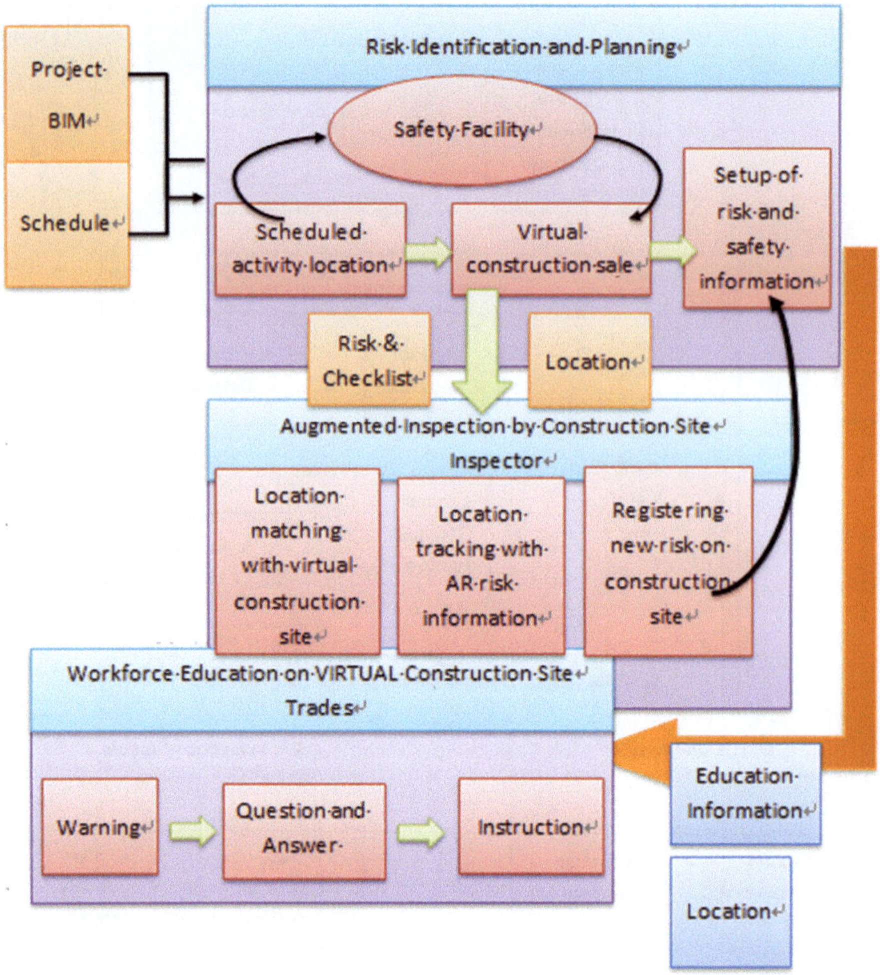

Fig. 10 Example of VR application with risk identification interface in construction industry

5 Cost–Benefit Analysis of VR Application in Construction Industry

5.1 Costs of VR

Devices in the past had a poor refresh rate, low resolution and field of view (FOV). In recent years, HMDs have been regaining popularity (Hilfert and König 2016). When users are using VR, the refresh rates of the display and updates may lead to motion sickness when users use it for a long period of time (Hilfert and König 2016). Nevertheless, technological breakthrough in recent years has alleviated the

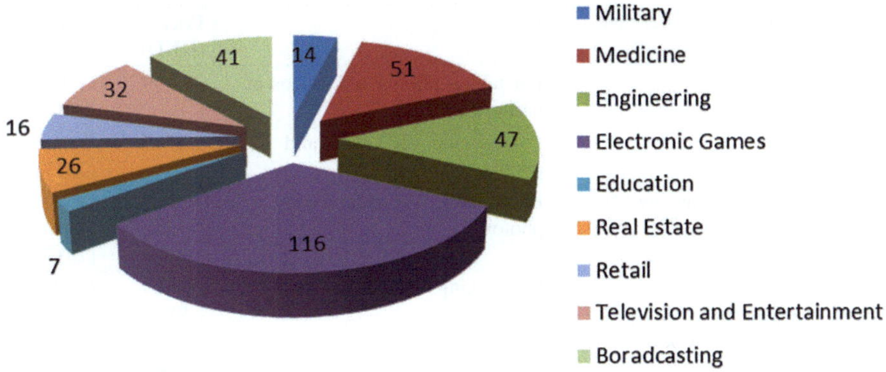

Fig. 11 Applications of VR in various different industries (Wong 2016) (author's figure)

abovementioned problems. Making its adoption for safety training feasible and affordable among relative wealthy construction companies.

In the construction industry, the 3D format is important for creating 3D building models. Time-consuming and costly real three-dimensional models motivate engineers to adopt VR. It is beneficial to the construction industry if engineers can minimise the complexity of VR presentations (Hilfert and König 2016).

Whilst safety training has always been considered one of the most important tools affecting workers' safety awareness and performance, the real construction work environment usually deviates from safety training and education, and it will be more difficult to identify safety risks and deliver the right information in the correct direction (Park and Kim 2013).

Having said that, however, there are some practical challenges for researchers in preparing imaginary VR content, e.g., 3D scenery, animations, pedagogical aspects, and the need to bring workers to a fixed facility (Sacks et al. 2013).

5.2 *Benefits of Adopting VR*

Compared to building real-life models on-site, simulation of a real-life situation is much cheaper (Hamari et al. 2016) (see the following table, which illustrates the tools for making VR videos). VR visualises the construction process by using two different construction methods. For example, it provides us with a deeper understanding of how bridges are built. The creation, deployment and interaction of 3D models over traditional practice through verbal description, pictures and diagrams are some of the major benefits of using VR (Hilfert and König 2016) (Fig. 11 and Table 1).

Tools for making VR heritage	Uses	Price (HKD)	Source
VR camera	For taking photos and videos of heritage sites	38,899	GoPro (2017)
VR gear	For viewing VR videos and photos	$599	Oculus (2017)
Viewing device: Samsung S6 4180 HKD	For viewing VR videos and photos		DC Denver (2017)
Oculus Rift VR Headset + Alienware Area-51 Signature Edition Desktop Bundle (Desktop) Acer Predator Z35 Monitor	For making VR Videos and photos	2699 USD US $899	Oculus (2017)
New Alienware 17 (Notebook) 15,992 HKD	For making VR videos on-site and seeing the results immediately after taking the video, etc.	1999 USD	Dell (2016a)
Cloud storage device Synology DiskStation DS216j 16 TB	Digital storage of VR videos and photos	$1400 HKD (about 175 USD)	
Double Robot, 9.7 in. iPad, 360 Panorama	Taking automated 360° VR videos steadily on the ground in the absence of cameraman obstruction	3499 USD	Double Robotics (2016)
9.7 in. iPad		7088 HKD	Apple (2016)
360 Panorama	Software for shooting VR videos	1.99 USD 16 HKD	i Tunes Review (2016)
DJI Inspire 1 RAW (Dual Remote), rapid mounting and dismounting	Taking videos on-site/outside and producing VR	41,839 HKD 5719 HKD	DJI (2017)
VideoStitch Studio v2 2288.70 HKD	Editing VR clips		

6 Mixed Research Method

In this research, we adopt a mixed research method with a combination of interviews, surveys and case studies. The research starts with interviews with construction practitioners and short online questions with construction practitioners/academics/tool providers, which ask their opinions with regard to the costs and benefits of VR in construction safety. It is widely believed that research triangulation shall increase the validity of the research (Li and Poon 2013a) and this method has also been applied in Li and Chau (2016).

In case studies, information can be gathered in Hong Kong and Seattle. It consists of interviews and official information collected from companies. One of them is one of the largest construction companies in Hong Kong and another one is a VR tool provider.

7 Construction Practitioners' Viewpoints on Virtual Reality

7.1 *Benefits of VR in Construction Safety*

7.1.1 Safety Training

A structural engineer who has 16 years of experience in Hong Kong suggests: "*VR is definitely useful, as it can make workers feel frightened and shocked when a simulated environment is shown to them. As far as I know, the Construction Industry Council (CIC) is using this technique for training*". A safety officer who works for a listed foundation company in Hong Kong comments: "*There is only one company which adopts VR training in the construction industry in Hong Kong. It has its own drawbacks and advantages. With regard to the advantages, we may not need to spoon-feed workers regarding safety and they may not be bored [by using VR]. Moreover, for some inexperienced workers, they can try more to prevent accidents from occurring*".

A top construction journal editor comments: "*VR connects the virtual world with the real world. VR can help train workers for training safety. Primarily on construction safety training only. We can put the sensor on the glove, touching the various 3D model*".

7.1.2 Improved Visualisation and Opportunity for Virtual Tour

A university professor in the United States suggests that adoption of VR may lead to a safer environment through improved visualisation. A construction practitioner who has worked in Las Vegas over the past 5 years suggests: "*The VR platform guarantees a real shortening of the needed time for training and, thus, a reduction of budget [for training], but the cost of equipment used for VR is very expensive*". Another researcher from the United States holds quite an optimistic viewpoint with regard to VR: "*Decisions made based on VR may lead to lower risks and a higher degree of certainty*".

An academic researcher in Australia suggests: "*It can help trainees raise awareness of general safety rules that a construction site cannot deliver. VR can provide a virtual tour that shows construction professionals 360 degrees of their virtual construction site*".

7.1.3 Generation Is an Important Factor Which Affects Their Likelihood to Accept the Technology

Previous research shows that generation is one of the important factors which affects individual's perception (Li 2015a), this is also reflected in our present study. Some of the interviewees think that the usefulness of VR in safety training depends on the age of the workers. A CEO of a refurbishment company suggests: "*Amid it being good for educating workers, the senior worker may not be interested in this education or management. The guidelines [laid down in virtual reality training] are strict and useful. The youth may show more interest in VR*".

A safety officer who works for a listed foundation company in Hong Kong states: "*To be honest, the idea or principle of VR is good, but will it be effective? In the long term, workers may find it boring to use... I watched 3D TV at the beginning; it was fantastic but I found it very boring towards the end.*"

7.2 Costs

Most of the interviewees agree that the costs of VR are the hardware and software. A VR tool provider in Fujian, China, suggests: "*It costs above 1000 RMB, not to mention the hardware and software costs, and I think the knowledge of their use and effectiveness is severely limited. Nevertheless, using virtual reality for construction safety training reduces the chance of losing life*".

A VR tool provider reveals that workable VR for the sake of safety education is not that cheap in Hong Kong: "*For VR in construction safety education, a display system is around $1 million and the VR device is around $0.5 million. To increase safety, it can imagine being able to walk to an aseptic area of construction. Moreover, VR can provide workers with a better understanding of the space and what a project will look like*".

A structural engineer who has 16 years of experience in Hong Kong suggests: "*I still doubt if the technique can make it come true when it is used popularly in construction, as the cost can be high. However, the cost must be lower when there are more companies applying the tool*".

Our research also reviews how younger officers are more positive in adopting VR in safety training. A safety officer who has worked for a listed foundation company in Hong Kong for two years suggests: "*If VR can show welding, reminding workers not to forget all the steps they should follow and to avoid explosions, then this unforgettable movement for them is better than the movement (explosion) that occurred. It's great. But I also agree that we should invest lots of money and time to develop this. We should let all workers use VR, but not just one or two workers, as effectiveness will be bad in the latter case. Nevertheless, will companies put lots of resources into developing it?*"

An officer from the International Labour Office in Switzerland suggests: "*VR is used to deal with virtual mapping on spaces. When they do this, they control a construction site easier. Somebody has immediate information immediately. Nevertheless, these are experimental technologies. He has no idea of how costly they are*".

A construction practitioner who has worked for over a decade in Iran and now works in one of the best universities in the United States also shares the same concern about costs: "*It needs infrastructure which is expensive. It needs professionals and specific expertise. The benefit is that as training could be improved, which leads to low risks and a higher degree of certainty*".

7.3 "I Do not Know What It Is" Is the Major Hang-up in Adopting VR On-site

Our research, to a certain extent, uncovers the reason why so many firms do not go out on a limb to act as a pioneer in virtual reality: quite a number of practitioners do not know what virtual reality is. We usually need to give them an example first to put them straight. Most of the interviewees will then be aware that VR refers to that which we can see in computer games. For example, a safety officer who has worked for seven years in Macau states: "*It is good that we can use VR. But can it really be used in construction works? Or can it only be used by architects? VR is usually used by designers only but not for construction works. BIM can show the working methods in different stages. I think VR is similar because I haven't seen that actually*".

Another architect shares similar comments based on his wild guess: "*The construction processes could be rehearsed with virtual reality. But I have no experience of this*".

8 Case Studies

8.1 Case Study One: VR Application in Safety Training in Hong Kong

In Hong Kong, one large-scale construction company has set up its VR training in a factory building. The VR program was designed by an overseas professor. In two-dimensional VR, users are immersed in a construction site environment with a simulated two-minute real-life cartoon through VR Cartoon, where they walk around construction sites.

Other three-dimensional VR simulated the "fall from height" situation. One of the research assistants said that he was frightened by the simulated "fall from height" VR clips. To a certain extent, that can allow learners, especially those with shadow experience, to have a deeper memory of the fall from height, leading to better safety awareness and, eventually, safety prevention. Nevertheless, only a few workers can use the VR techniques, as the training time is limited. Most of these lucky ones are new comers in the construction industry (Fig. 12).

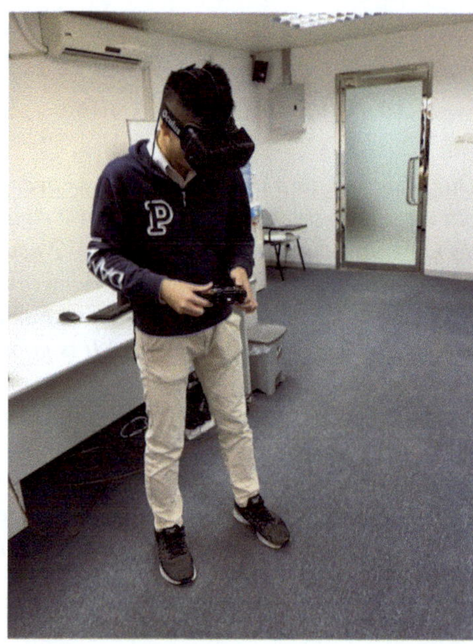

Fig. 12 Our research assistant engaged in safety training with VR (author's photo)

8.2 Case Study Two: VR Application in Planning Stage in Seattle, United States

VR allows different stakeholders on the same page. For example, Visual Vocal, a firm in Seattle, has developed a multiuser collaborative VR system. A project team can utilise their online V|V system to make a three-dimensional model, which can be shared with other stakeholders via their special app; then, viewers can view the 360° design via a binocular viewer, moving their head accordingly (Visual Vocal 2017).

The following pictures are shown in a VR program developed especially for a building in its planning stage. Different stakeholders can view the VR clips and choose which types of building structures they prefer most by clicking on multiple-choice questions A, B and C. A VR survey, as such, allows the architecture company to make the right decisions that fit most of the stakeholders' needs and wants inside the VR meeting simultaneously or at different times and locations. When most of the participants saw these VR tools in Seattle, they were very excited by the tools and believe that they could help them to make choices for their construction projects. It may improve safety as well, as they can have a better understanding of the projects before they build (Figs. 13, 14 and 15).

Fig. 13 Different structural options are available in the VR clips (example one) (author's photo)

Fig. 14 Different structural options are available in the VR clips (example one) (author's photo)

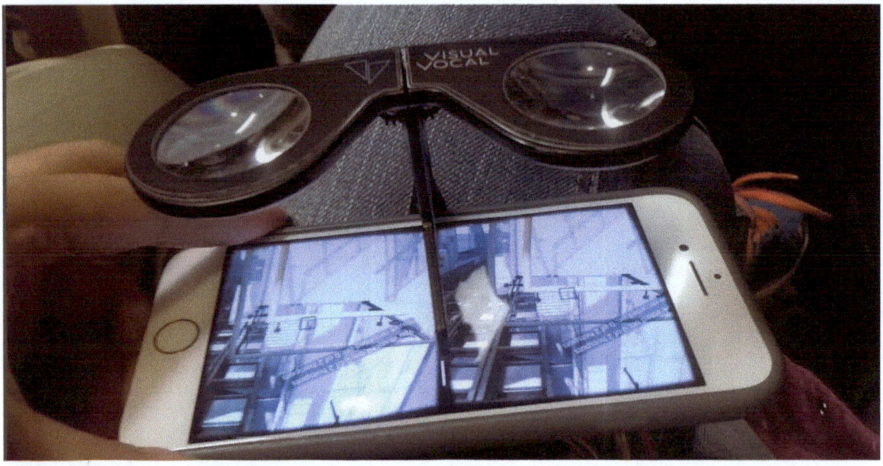

Fig. 15 The special lens designed by Visual Vocal (author's photo)

9 Conclusion

VR simulates real-world situations and allows users to feel as if they are there. It is now widely applied in shopping mall promotions, driving improvement and the education sector. The construction industry has also adopted VR in improving safety awareness. A simulated fall from height creates a scary environment for construction industry newcomers. Apart from that, a simulated building structure before it is constructed allows stakeholders to make decisions online. That essentially saves many costs as compared to building a real model for discussions.

Acknowledgements The author would like to thank Peter Koon's contribution in this chapter.

References

Apple. *Compare the models of different iPADs* 2016, cited 10 November 2016. Available from http://www.apple.com/hk/ipad/compare/.

Artaud, Antonin. 1938. *Le Théâtre et son double*. France: PgW.

Azhar, Salman, and Rafiq M. Choudhry. 2016. Capacity building in construction health and safety research, education, and practice in Pakistan. *Built Environment Project and Asset Management no.* 6 (1): 92–105.

Bailenson, Jeremy N., Jim Blascovich, and E. Rosanna. 2008. GuadagnoSelf-Representations in Immersive Virtual Environments. *Journal of Applied Social Psychology* 38: 2673–2690.

Bong, Shanan, Raufdeen Rameezdeen, Jian Zuo, Rita Yi Man Li, and Gui Ye. 2015. The designer's role in workplace health and safety in Construction Industry: post-harmonized regulations in South Australia. *International Journal of Construction Management no.* 15 (4): 276–287.

DC Denver. *Samsung Galaxy s6* 2017, cited 2 April 2017. Available from http://www.dcfever.com/phones/specification.php?id=2350.

de Vries, B., S. Verhagen, and A.J. Jessurun. 2004. Building management simulation centre. *Automation in Construction* 13: 679–87.

Dell. *New Alienware 17 Gaming Laptop* 2016, cited 10 November 2016. Available from http://www.dell.com/en-us/shop/productdetails/alienware-17-laptop.

DJI. *DJI Store* 2017, cited 2 April 2017. Available from https://m.dji.com/product/inspire-1-raw?from=menu_products.

Double Robotics. *Robot Double* 2016, cited 10 December 2016. Available from https://secure.doublerobotics.com/buy?numberOf360Sets=1.

Google. *Virtual Reality* 2016, cited 15 April 2016. Available from https://www.google.com/trends/explore#q=virtual%20reality.

GoPro. *Advanced Systems* 2017, cited 2 April 2017. Available from http://zh.shop.gopro.com/APAC/virtualreality/omni—all-inclusive/MHDHX-006-master.html.

Grant, Aneurin, and Jimmie Hinze. 2014. Construction worker fatalities related to trusses: An analysis of the OSHA fatality and catastrophic incident database. *Safety Science* 65: 54–62.

Hilfert, Thomas, and Markus König. 2016. Low-cost virtual reality environment for engineering and construction. *Visualization in Engineering* 4 (2): 1–18.

Hsu, Kuei Shu. 2016. Application of the environmental sensation learning vehicle simulation platform in virtual reality. *Eurasia Journal of Mathematics, Science & Technology Education* 12 (5): 1477–1485.

i Tunes Review. *360 Panorama* 2016, cited 10 November 2016. Available from https://itunes.apple.com/us/app/360-panorama/id377342622?mt=8.

Hamari, Juho, David J. Shernoff, Elizabeth Rowe, Brianno Coller, Jodi Asbell-Clarke, and Teon Edwards. 2016. Challenging games help students learn: An empirical study on engagement, flow and immersion in game-based learning. *Computers in Human Behavior no.* 54: 170–179.

Li, Rita Yi Man. 2015a. *Construction safety and waste management: An economic analysis.* Germany: Springer.

Li, Rita Yi Man. 2015b. Generation X and Y's demand for homeownership in Hong Kong. *Pacific Rim Real Estate Journal* 21 (1): 15–36.

Li, Rita Yi Man. 2016. *VR as an edutainment tool in secondary school sustainable development learning workshop*. Hong Kong: In Hong Kong Shue Yan University.

Li, Rita Yi Man 2017, forthcoming. 5D GIS Virtual Heritage. *Procedia Computer Science* no. 103.

Li, Rita Yi Man, and Kwong Wing Chau. 2016. *Econometric analyses of international housing markets*. UK: Routledge.

Li, Rita Yi Man, Kwong Wing Chau, and Daniel Chi Wing Ho. 2017, forthcoming. "Dynamic panel analysis on construction accidents in Hong Kong." *Asian Journal of Law and Economics*.

Li, Rita Yi Man, Herru Ching Man, Yu. Li, Cho Kei Mak, and Tang Beiqi Tang. 2016. Sustainable smart home and home automation: Big data analytics approach. *International Journal of Smart Home no.* 10 (8): 177–187.

Li, Rita Yi Man, and Sun Wah Poon. 2013a. *Construction safety*. Berlin: Springer.

Li, Rita Yi Man, and Sun Wah Poon. 2013b. Demand for construction safety. In *Construction safety*, ed. Rita Yi Man Li, and Sun Wah Poon. Germany: Springer.

Li, Rita Yi Man, Jian Zuo, and Terdoo Fanen. 2015. Supply and demand of construction safety regulations in developing countries. In *Construction Safety and Waste Management: An Economic Analysis*, 93–109. Cham: Springer International Publishing.

Lin, Jih-Hsuan Tammy. 2017. Fear in virtual reality (VR): Fear elements, coping reactions, immediate and next-day fright responses toward a survival horror zombie virtual reality game. *Computers in Human Behavior no.* 72: 350–361. doi:10.1016/j.chb.2017.02.057.

Lingard, Helen, Matthew Hallowell, Rico Salas, and Payam Pirzadeh. 2017. Leading or Lagging? Temporal analysis of safety indicators on a large infrastructure project. *Safety Science* 91: 206–220.

Lucas, J., Walid Thabet, and P. Worlikar. 2008. A VR-based training program for conveyor belt safety. *Journal of Information Technology in Construction* 13: 381–407.

Oculus. *Rift+ Touch* 2017, cited 2 April 2017. Available from https://www.oculus.com/rift/.

Park, Chan Sik, and Hyeon Jin Kim. 2013. A framework for construction safety management and visualization system. *Automation in Construction* 33: 95–103.

Sacks, Rafael, Amotz Perlman, and Ronen Barak. 2013. Construction safety training using immersive virtual reality. *Construction Management and Economics* 31 (9): 1005–1017.

Samsung. *Samsung brings the world of mobile virtual reality to Australia with the Gear VR* 2014, cited 3 April 2017. Available from http://www.samsung.com/au/news/local/samsung-brings-the-world-of-mobile-virtual-reality-to-australia-with-the-gear-vr.

Visual Vocal. *How will virtual and augmented reality make the 3D design process more Enjoyable?* 2017, cited 23 April 2017. Available from http://www.visualvocal.com/.

Wong, Lam Jeng. *International Real Estate Investment Heat: the Rise of VR Application* 2016, cited 18 September 2016. Available from http://www.digitimes.com.tw/tw/rpt/rpt_show.asp?cnlid=3&v=20160421-99&n=1&wpidx=8#ixzz48KALKvRv.

Zhao, Dong, and Jason Lucas. 2015. Virtual reality simulation for construction safety promotion. *International Journal of Injury Control and Safety Promotion* 22 (1): 57–67.

Chapter 7
Smart Working Environments Using the Internet of Things and Construction Site Safety

Abstract Safety supervision on construction sites has always been undertaken with the help of safety officers and other safety personnel. In recent years, smart working environments using the Internet of Things have turned a new page in terms of safety management on construction sites. Direct communication between different "smart things" such as sensors, smart phones and databases offers a new angle on safety management. In this chapter, first an introduction to the Internet of Things is provided. This is followed by a study on the popularity of the IoT in different countries and cities—undertaken with the help of big data provided by Google. After this, the results of an interview concerning real-life applications of IoT in Australia and the proposed IoT application of IoT to road repair works will be presented.

Keywords Internet of things · Big data · Construction safety

1 Introduction

The construction industry is one of the most hazardous among all the industries which have led to large numbers of fatalities around the globe (Fig. 1). Japan, Russian Federation and Colombia record the largest number of accidents which have happened on construction sites with 327, 255 and 195 fatalities, respectively, in 2015 (International Labour Office 2017; Li and Poon 2011, 2013). Previous research shows that about 90% of workplace injuries occur due to unsafe work behaviours and practices; proper safety management is important to lower risks that arise in unsafe situations and activities. Nowadays, Smart Work Environments (SWE) monitor activities on-site: in terms of workers, machinery and tools. Such environments provide an infrastructure which interconnects the virtual and physical "things", with the help of the IoT. Semantic services and the service-oriented architecture concept provide decentralised architectures to facilitate IoT Services

R.Y.M. Li, *An Economic Analysis on Automated Construction Safety*,
DOI 10.1007/978-981-10-5771-7_7

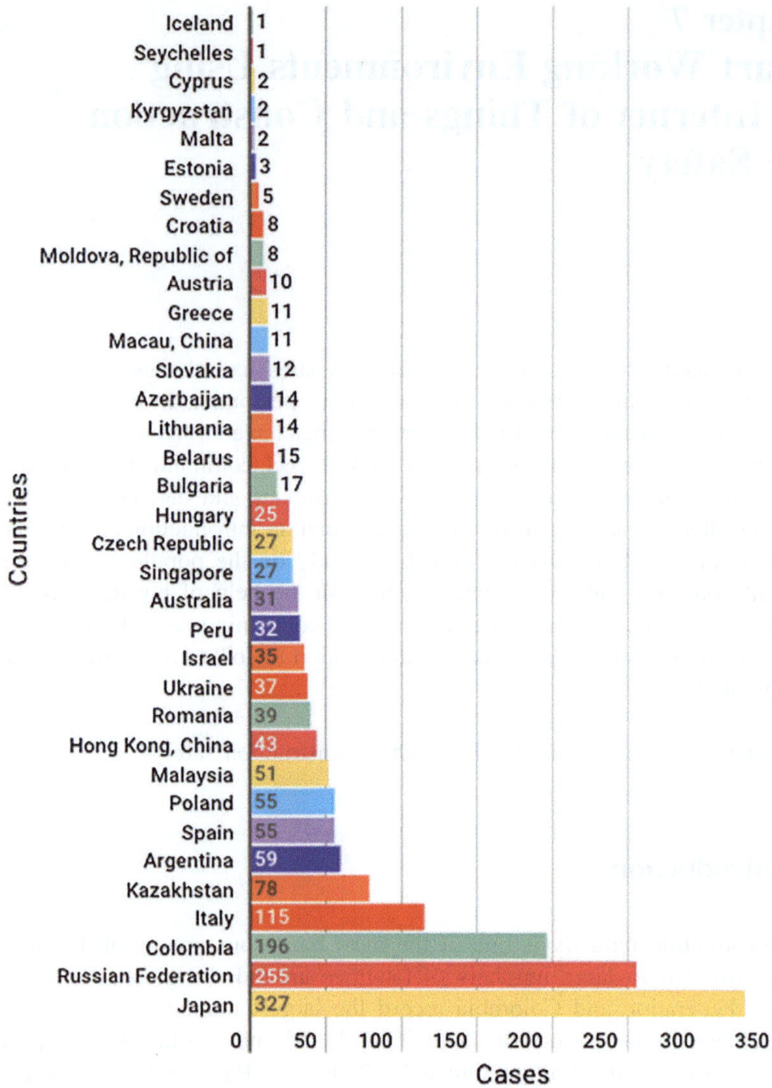

Fig. 1 The number of construction-related work fatalities, in 2015, worldwide (International Labour Office 2017)

and smart objects interaction (Teimourikia and Fugini 2017) (Fig. 2). Because it combines smart objects, or "things", such as actuators, RFID tags, mobile phones and sensors, IoT is a new generation technology which makes intelligent processing and comprehensive sensing possible (Ding et al. 2013).

2 Internet of Things (IoT) and Smart Object Interactions

Despite the fact that the IoT occupies a small part of the ICT market, it provides a good foundation for future global technological developments. It can be deployed in many fields to improve people's quality of life, such as workplace and home support, health care, product management, transportation logistics, environmental monitoring, inventory, surveillance, and security (Ding et al. 2013). Figure 3 shows the four leading industries that have adopted IoT technologies. A recent survey conducted by Enterprise Management Associates with 351 organisations' showed that 47% of them consider IoT as a technology crucial to their business (Quest 2017).

IoT middleware platforms allow consumers to retrieve data without knowing the full technical details of the IoT resources involved: the data processing components and sensors. Nevertheless, data retrieval and IoT middleware platform configuration impose a significant challenge for consumers as these activities require domain expertise. The context-aware sensor configuration model simplifies the IoT middleware platform configuration process, such that non-technical personnel can retrieve data easily (Perera and Vasilakos 2016). Figure 4 illustrates the requirements that the IoT must accord with.

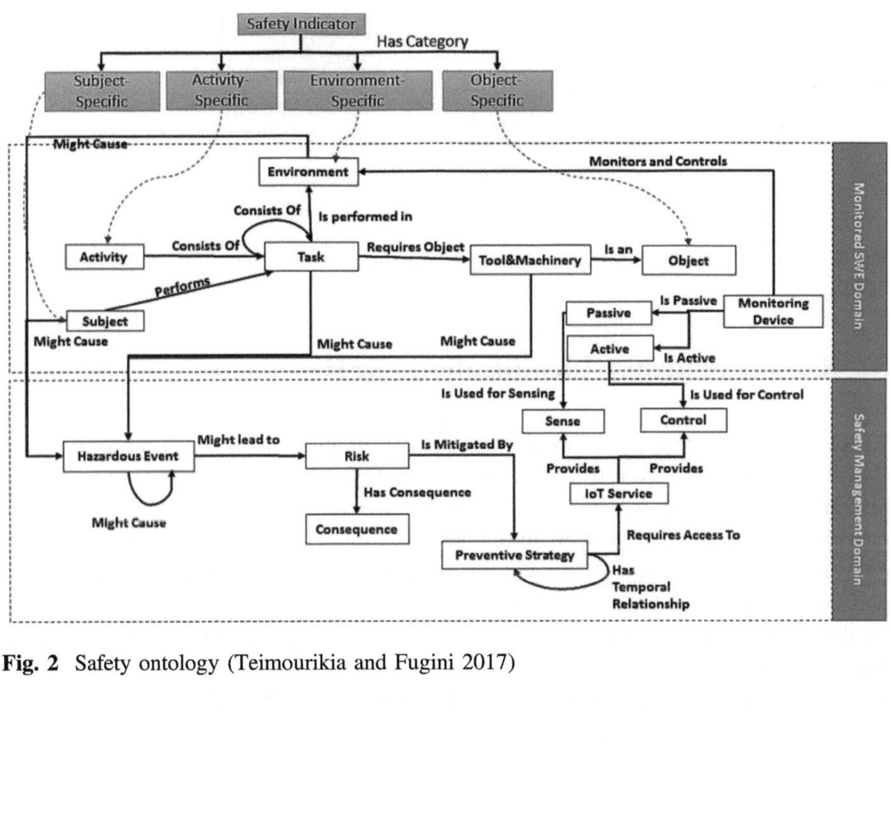

Fig. 2 Safety ontology (Teimourikia and Fugini 2017)

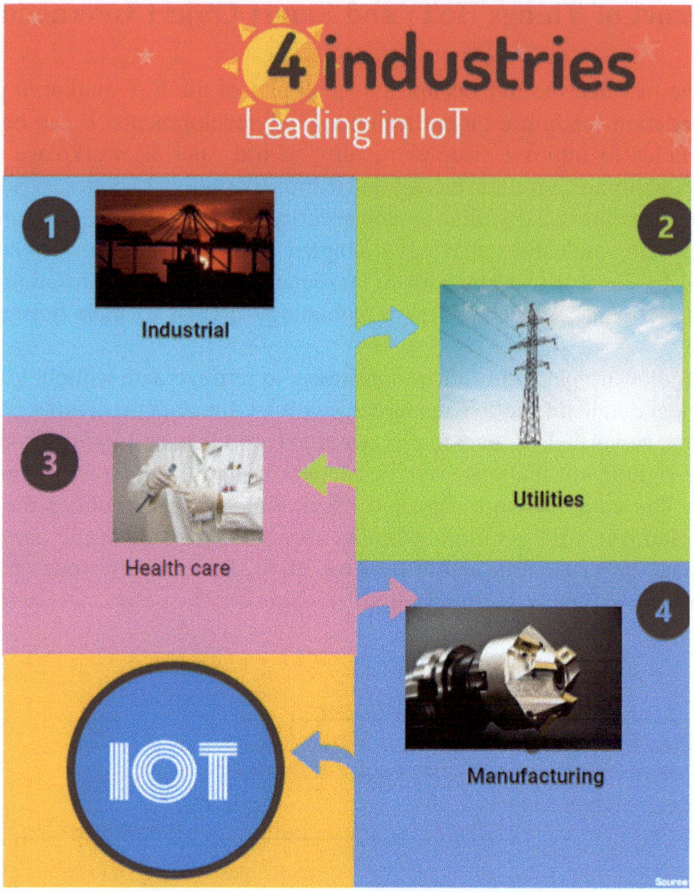

Fig. 3 Four major industries that have adopted the Internet of Things (Quest 2017) (author's figure)

3 Radio-Frequency Identification (RFID)

RFID's ability to perform as an auto-identification technology was first used by the Royal Air Force in World War II to differentiate between enemy and friendly aircraft. Friendly planes were equipped with "active", RFID tags. These tags were energised by an attached power supply and interrogated by an RFID reader. Modern day applications rely on a similar kind of communication between RFID readers and tags. Nowadays, tags are powered by an electromagnetic field which is emitted by the reader. Radio signals inform any reader in close proximity of the tag's serial number—which identifies any item bearing the tag. Smart tags are used to track objects and send safety alert (Angell and Kietzmann 2006) (Table 1).

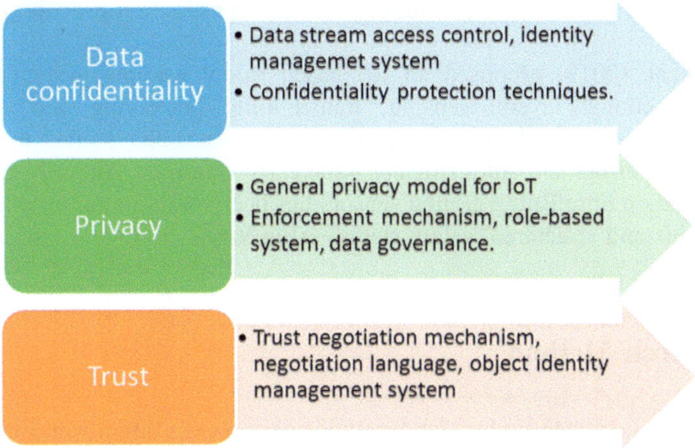

Fig. 4 Requirements for the IoT (Miorandi et al. 2012) (author's figure)

Table 1 Safety alert (Angell and Kietzmann 2006)

Types of reminders	Safety alerts and warnings
Construction workers' reminders	• Beware of dangerous materials • Beware of electric box • Beware of fall from height • Beware of the slippery floor • Beware of the vehicles
Reminder for equipment operator	• Beware of the edge • Beware of workers • Don't drive outside the site

4 IoT Application on Construction Sites

The IoT allows for automatic responses in real time. With regards to safety management on-site, the rise of the IoT will result in a paradigm shift: from a reactive manual approach to proactive automated methods. Safety management systems can be designed and built in response to specific requirements according to a construction projects' characteristics (Ding et al. 2013).

An IoT-based safety early warning system is a highly dynamic, distributed, networked system consisting of many devices that gather, investigate and share safety information. Sensor interconnection monitors, RFID devices, construction participants, and the network create an intelligent loop for safety prediction, monitoring and control. An IoT-based safety early warning system can be developed with the integration of smart sensor technologies, such as piezoelectric sensors and FBG sensors and location tracking technology, such as WSN and/or RFID. These technologies can share safety information in real time and monitor workers and “things” on construction sites throughout the construction process (Ding et al. 2013).

Ding et al. (2013) proposed an IoT-based safety warning system with four different functional layers: sensing, transmitting, perceiving and controlling (Fig. 5). An RFID-based labour tracking system was implemented to track and reduce human errors and automatically issue early warnings. An SHM monitoring system based on fibre Bragg grating sensors has been used to monitor underground environments and structures’ safety statuses (Fig. 6).

5 Research Method

Included in this chapter are big data analytics performed using a worldwide google search, a content analysis of previous literature, interviews, a case study of a construction site in Adelaide, and finally a conceptual proposal regarding the adoption of IoT to enhance road safety construction.

5.1 Big Data Analytics

Data analytics is the use of data collection and analysis to optimise decision-making. By this methodology, researchers can review large amounts of raw and unorganised data and identify patterns which can help decision-makers or organisations to better understand human behaviours. It is a useful tool for gaining meaningful insights into how sentiments, information, and opinions are generated. Moreover, such analysis determines prevailing issues in the market. Organisations nowadays often use data analytics to infer business intelligence from social media content in order to make decisions or to formulate various business strategies for review by decision-makers. Data analytics provides opportunities for business intelligence, creating a platform whereby the decision-makers have a better understanding of a business’s performance. Business intelligence encompasses the processes and technologies used to obtain timely and valuable insights into a business. However, data analytics and BI are only an aid for decision-makers; they should always analyse based on their own analytical skills after obtaining the results from a data analysis and not blindly make their decision according to the results from this. There are challenges related to data analysis, including the availability, integrity and relevance of the data. Many managers and organisations do not have the ability to capture data; they may lack, for instance, computer skills and knowledge, or the data they have may contain noise. Even if the data is readily provided and the decision-makers have full access, they have to consider the completeness and quality of the data (Li et al. 2016).

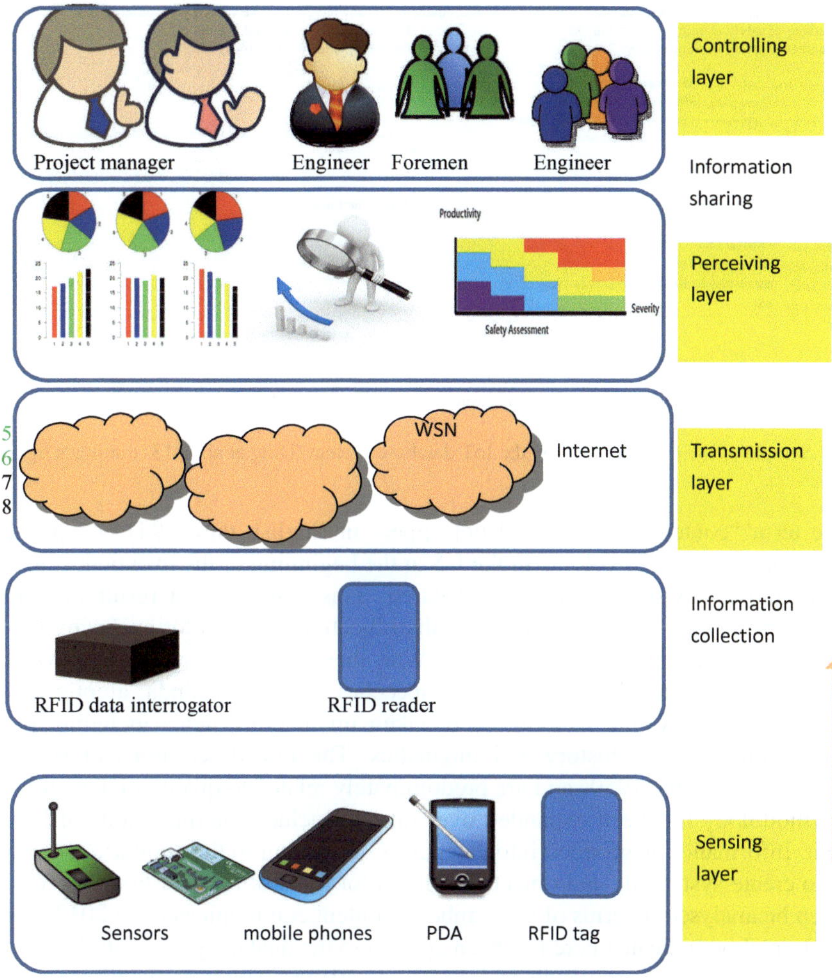

Fig. 5 An IoT-based safety warning system (Ding et al. 2013) (author's figure)

5.2 *Content Analysis*

Content analysis has always, since around 1941, been used to investigate and describe data by extracting and evaluating, in a systematic way, the occurrence of latent content in a body of textual material. An appropriate content classification scheme is an important first step. Content analysis, like any other (quantitative) research method, has its weaknesses and strengths. To successfully develop a content analysis, classification and/or measurement of the data must be undertaken with objectivity, rigour and exactness; this requires a large degree of personal judgment and an in-depth knowledge of the subject matter (Li and Li 2013).

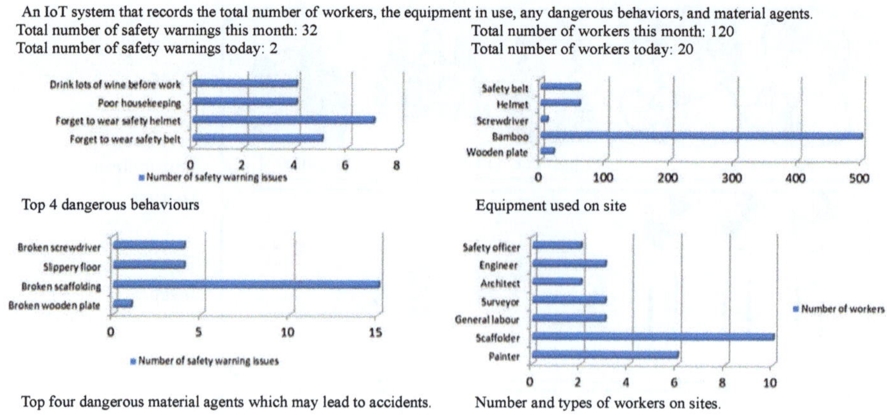

Fig. 6 Sample of the data recorded by the IoT database system (Ding et al. 2013) (author's figure)

The term "content analysis" did not appear in English until 1941. As printed mass media increased in volume in the US at the beginning of the twentieth century so quantitative newspaper analysis appeared. This was the end result of many efforts to create easy to use and scientifically objective methods for analysing news articles. At the end of the Second World War, these methods had become widely used to study journalistic texts, political speeches, and propaganda, among other things. Subsequently, the methods were taken up in other fields including psychology, anthropology, history and linguistics. The first descriptions of content analysis date from the 1950s and are predominately related to quantitative methods; the methodology has been expanded over time to include interpretations of latent content. Information is grouped into categories, according to the similarities found, so as to create systematic and objective criteria for transforming written text which can then be analysed in terms of the symbolic content communicated (Li 2013). The research method indicated here has been applied to the analysing of safety literature related to the construction industry (Li et al. 2015b), smart-home research (Li 2012), construction waste management (Li et al. 2015a) and so on.

5.3 *Interviews, a Real-Life Application of an IoT Application in Adelaide, and a Proposal for an IoT Application*

We have conducted interviews[1] with construction practitioners, tool providers and academics in order to get to know more about their viewpoints with regard to IoT's

[1]Overseas participants may fill in the forms only which have the same set of questions as that contains in the interviews.

costs and benefits in relation to construction safety. We have also used a case study to show that RFID has been applied on a construction site in Adelaide.

6 Results

6.1 *The Trend Towards IoT in Recent Years: A Big Data Analytics Approach*

The rapid growth of the mobile internet in recent years has led to a development in the use of the IoT. To study the popularity of the IoT, we adopted a big data analytics approach which had previously been used by Li et al. (2016). Figure 7 illustrates that there is a rapid growth in the number of Google searches relating to IoT in the last two years. Singapore, India, South Korea and Hong Kong have been the source of the largest number of searches of "IoT" in Google, according to Fig. 8. With regard to the cities which have sourced the largest number of such searches, these include San Jose, Chennai and Hyderbad (Fig. 9).

As the number of related searches in search engines such as Google have a direct relationship with people's interests in a topic, the above is sufficient evidence to conclude that IoT not only attracts the interest of people in developed countries but also of those in developing countries such as India. Cities which are well-known for their activities relating to information technology record the largest number of searches, irrespective of their economic status. For example, Bangalore, Chennai and Hyderbad are located in India, and are well-known for information technology (Fig. 5). The top five topics which are searched for alongside IoT, however, are not related to the construction industry. They are "internet of Things 2016", "raspberry pi internet of things", "raspberry pi", "nest and iot devices". This, to a certain extent, shows that those who are aware of the existence of the IoT mainly come from the information technology industry. Thus, there is a great deal of room for improvement as regards the construction industry.

6.2 *Costs and Benefits of IoT, According to the Literature: Content Analysis Results*

Previous research indicates that due to poor communication between the "things" (Podgórski et al. 2017) and project complexity (Ding et al. 2013), IoT is not popular on sites. Nevertheless, it has potential to be widely used in the future. Tools such as RFID can used used for safety alert which reduce the likelihood of accidents. Table 2 lists the costs and benefits of IoT according to previous literature.

Fig. 7 The increase in the number of searches for Internet of Things in Google, from 2004 to April 2016 (Google 2017)

Fig. 8 Countries that have generated the most Google searches for Internet of Things, from 2004 to 2016 (Google 2017)

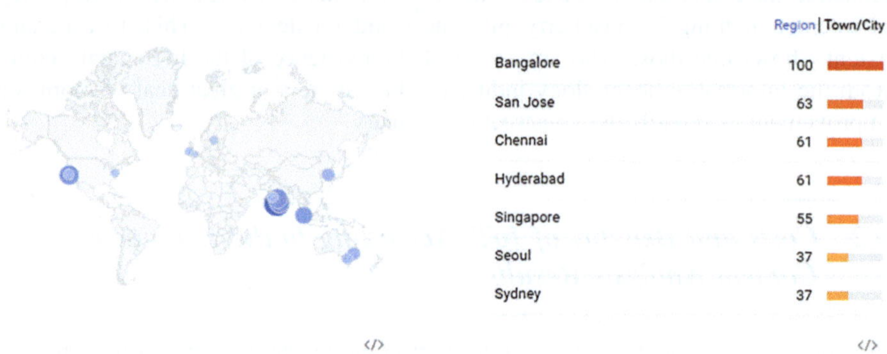

Fig. 9 Cities that are the most interested in the Internet of Things (Google 2017)

Table 2 Results of literature review on the costs and benefits of IoT in the construction industry

Author	Costs	Benefits
Construction		
Ding et al. (2013)	Due to the complexity of construction activities and the dynamic nature of construction resources and environments IoT technology is less developed for large scale construction projects. The future smart construction site would greatly benefit from the adoption of IoT to establish new approaches for construction activities. By combining smart objects or "things", such as RFID tags, actuators, mobile phones and sensors, IoT allows intelligent processing and comprehensive sensing (Ding et al. 2013)	Working for underground construction projects is high risk due to unpredictable geological and hydrological conditions, concurrent activities on-site and the dynamic nature of construction sites using different construction methods. RFID-based labour tracking systems can act as a proactive safety precaution approach which could be used for the prevention of human error and behavioural risks as well as for dynamic identification. It is particularly important to underground construction due to complex interactions between the vast amount of different resources and pace constraints
Fang and Song (2011)		Internet of Things technologies and data mining can be put to use in a wide range of applications such as architecture, bridge, collapse column, ground facilities, roads, and underground coal strata, water, gas and so on. It enhances production safety and can create a comfortable and safe working environment
Other industries		
Zhao et al. (2013)		IoT techniques are used to monitor atmosphere, water, sound, soil, and wind environmental indicators to implement collection, transportation, early warning, forecasting, and modelling, and the application of environmental information using online, real-time, in situ and long-distance approaches
Podgórski et al. (2017)	Only a small amount of potential IoT applications are available at present, as objects covered by these applications usually communicate poorly with one another and are equipped with primitive intelligence	It is easy to install, maintain and very flexible. There is a strong potential to advance management processes by providing real-time access for workers

(continued)

Table 2 (continued)

Author	Costs	Benefits
Balaji and Roy (2017)		It allows for greater accessibility, ubiquitous connectedness, interactivity, integration into the physical environment, real-time synchronisation, localised and personalised information using context-awareness, increase support and greater monitoring

6.3 *Results of the Interviews*

Seven interviewees from Hong Kong, three from the United States, three from mainland China, two from Australia and one from Switzerland took part in the interviews. They were from different contractors, government department, safety consultancy firms, small-scale refurbishment company, universities, labour union and only one innovative manager knew the terminology of IoT well and asked for clarification on the exact type of IoT that we were referring to. Nevertheless, in spite of the fact that most of the respondents were not knowledgeable about the terminology, many of them did realise that the IoT is used in the construction industry. Many of them pointed out that at least one of the companies has installed a computer chip in the workers' helmets. When a worker is inactive for a long time, a warning signal will wake up the worker. When they are told what IoT is—for instance, given the example of the chip installed in the helmet—most of the safety officers agreed that IoT can improve construction safety. They were also aware that QR codes and RFID are used on construction sites.

6.3.1 Benefits

An experienced construction practitioner in Shenzhen suggested that the benefit was that it means greater safety for employees and for the project itself. A fifty-year-old section chief in a construction company in Fujian, China, stated that IoT requires the proximity and availability of computer devices in relation to workers. This can be beneficial for various actors and from core business perspectives: from lean construction management to the management of the entire building lifecycle; also, relative to these benefits, the price is not high. A researcher in Melbourne has suggested that the use of IoT has the benefit of enhancing job and site safety—procedures and reports are accessible in all locations. For example, drivers can be required to wear an activity band so that this device can alert the driver when it detects that she/he is falling asleep. A tool provider in Hong Kong suggests that even though there are costs in relation to these technologies, their use can allow for the integration with further applications like smart houses or smart buildings or even smart cities.

A researcher in Switzerland highlighted the fact that the Internet of Things speeds things up. Whilst IoT and BIM speed up the sharing of knowledge, the IoT allows us to demonstrate how to digitalise on-site. The workers can more easily ascertain who are doing what jobs. Where there are risks, all the relevant information is collected automatically. The use of the IoT can speed up all the on-site processes. It can speed up the acquiring of data. Dangerous locations can be avoided. Tags can be used to store some of the safety information, the locations of workers, etc.

A university professor from the United States commented that "*IoT allows several building structure to exchange data for monitoring the structure, monitor for safety of workers. More data can be collected in real time. You can use these IoT inform the supervisors of what is going on and may be automated warning when the worker is entering to the dangerous zone.*"

6.3.2 The Drawbacks of the IoT

Although an innovative manager, who was an interviewee, pointed out that the IoT can, of course, improve construction safety by sending warnings to the workers, nevertheless, if you are worker, you may not want your manager to know whether you are working or having a rest. Nobody wants to be kept watch over when they are at work; there are concerns about privacy. Hence, even though the information technology department and the senior manager may want to use this technology or something similar, many workers would be strongly opposed to its use.

A professor with over thirty years of experience shares similar thought that "there are various reasons [for not adopting IoT on sites]: invasion of privacy, workers don't want to be monitored 24/7. It is difficult to create a coherent network. It is difficult to have power for the sensors. It is not easy to find the right source of power. Some have been using solar power."

In view of this, whether or not the workers have a sufficient level of information technology at their disposal is not the crux of issues which lead to the non-adoption of IoT. Whilst somebody may criticise that workers may not be well-educated and do not use high-tech stuff, when we look at construction sites in Hong Kong, there are many workers who use mobile apps; almost everyone is using them. That actually provides some clue with regard to workers' capability and possibility to learn and use some new information technology tools. Another construction practitioner argued that the cost of using IoT is that workers might be placed in danger of losing their jobs as some of the supervision jobs could potentially be replaced by the IoT system. The use of IoT can stop the haemorrhaging of money on supplies that get mislaid during a project, and on fuel costs related to all those large items of construction equipment.

6.4 RFID Application in Adelaide

At the Royal Adelaide Hospital, the information technology manager mentioned that they used RFID on their construction site in order to enhance site safety. Each of the large window panels has an RFID tag containing information such as the position of where the window should be installed. This arrangement can enhance site safety and save costs as an incorrect installation may lead to more work and increase the chances of accidents.

6.5 Other Possible Application of IoT on Construction Sites

A chairman of the labour union in Hong Kong suggested that the Internet of Things can be applied on-site to alert the construction workers to safety risks and hazards. Based on his suggestion, we have made our conjecture in Li (2017, forthcoming-b). For example, when the weather is extremely hot and workers work for over three hours, a message can be sent to the workers to prompt them to drink more water. When there is high humidity or rainfall, the system can send alerts the workers so that they will be more aware of the wet and slippery surfaces. Likewise, when too many workers work on-site, the system can generate a "no entry" instruction. All these things can be done using straightforward computer programming based on Python (Figs. 10 and 11) (Li 2017, forthcoming-b).

Another important concern is the relatively high risk to workers who set up road blocks. Robots and robotic arms can help in this case. Instead of humans placing all the cones individually, robots can be used to place all the cones, so that the likelihood of accidents involving injury or death will be substantially reduced (Fig. 12). Even though there is a concern regarding the relatively high price of using robots/robotic arms, skyrocketing labour costs and the continuing reduction in robot and robotic arm prices suggest that this may be a cost-effective choice.

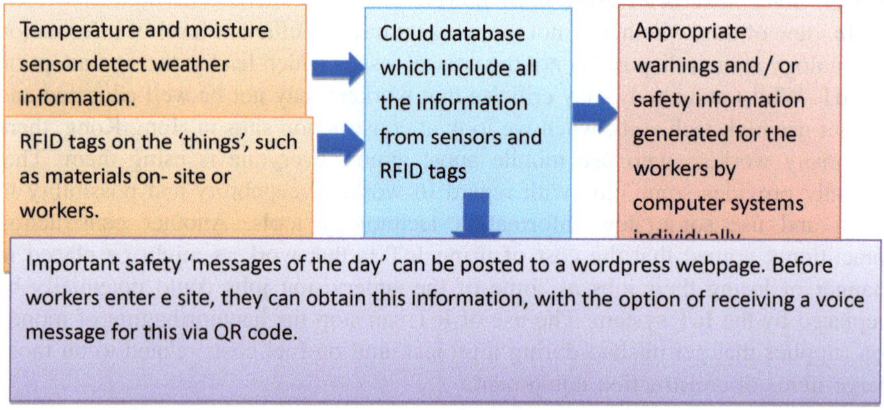

Fig. 10 IoT-based safety information sharing system (author's figure) (Li 2017 forthcoming)

```
import os
temperature = 40
hours = 4
if (temperature>31 degree celcius & working hours>=3):
print ("Please drink water.")
os.system ("start C:\Users\Downloads\hello.wav")
if (rainfall == 1) or (humidity > 90):
print ("Beware of the slippery floor")
os.system("start C:\Users\Downloads\hello.wav")
if (number of workers> 4):
print ("No entry is allowed")
os.system("start C:\Users\Downloads\hello.wav")
```

Fig. 11 Sample programming for road construction safety warning

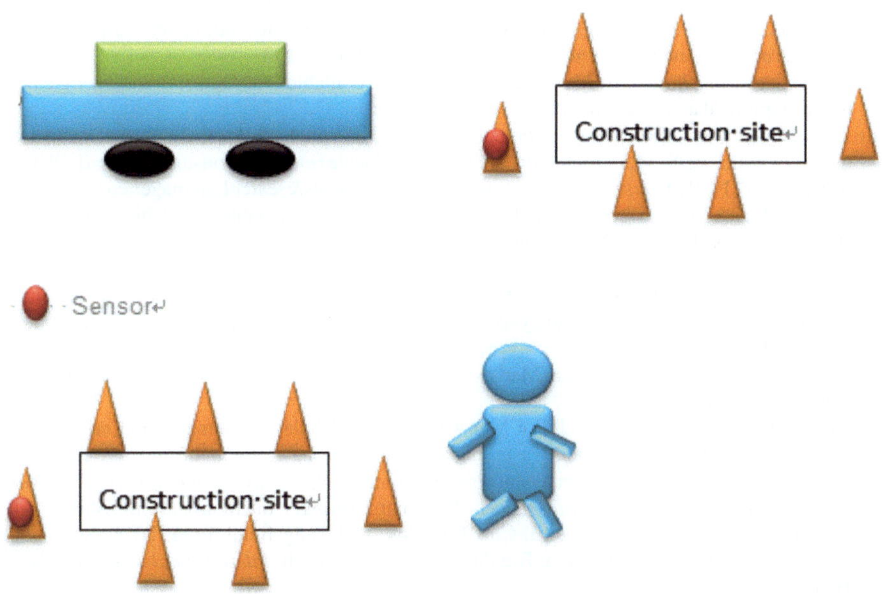

Fig. 12 Sensor and construction site accidents prevention in road construction and/or repair works (Li 2017 forthcoming)

Consider the comparison between a robot costing HK$30000 and a worker's salary of HK$1500 per day, the robot represents a bargain if it works well throughout its service life. This approach could also be used to solve the problem of labour shortage in some places like Hong Kong (Li 2017, forthcoming-b).

7 Discussion and Conclusion

IoT which link all the "dead" things together via Wi-fi may be used on-site to alleviate the safety problems. One construction company in Hong Kong has adopted IoT for their work on-site in Hong Kong. Although it may enhance construction safety, workers may not wish to be watched so closely. In Australia, RFID, a type of IoT tool, is adopted on-site to ensure the window panels are installed in the right position so as to prevent redo and unnecessary safety risks.

References

Angell, Ian, and Jan Kietzmann. 2006. RFID and the end of cash. *Communications of the ACM no.* 49 (12): 91–96.

Balaji, M.S., and Sanjit Kumar Roy. 2017. Value co-creation with Internet of things technology in the retail industry. *Journal of Marketing Management no.* 33 (1): 7–31.

Ding, L.Y., C. Zhou, Q.X. Deng, H.B. Luo, X.W. Ye, Y.Q. Ni, and P. Guo. 2013. Eal-time safety early warning system for cross passage construction in Yangtze Riverbed Metro Tunnel based on the internet of things. *Automation in Construction* 36:25–37.

Fang, Liu Yu, and Jin Xing Song. 2011. Sing the internet of things technology constructing digital mine. *Procedia Environmental Sciences* 10 (Part B):1104–1108.

Google. *Internet of Things* 2017, cited 25 April 2017. Available from https://trends.google.com/trends/explore?date=all&q=internet%20of%20things.

International Labour Office. *Construction Accidents* 2017, cited 25 April 2017. Available from http://www.ilo.org/Search5/search.do?sitelang=en&locale=en_EN&consumercode=ILOHQ_STELLENT_PUBLIC&searchWhat=construction+accidents&searchLanguage=en.

Li, Rita Yi Man. 2012. A content analysis in Hong Kong sustainable smart home knowledge sharing via world wide web. *Communications in Computer and Information Science no.* 350: 265–269. doi:10.1007/978-3-642-35594-3_37.

Li, Rita Yi Man. 2013. The usage of automation system in smart home to provide a sustainable indoor environment: A content analysis in web 1.0. *International Journal of Smart Home no.* 7 (4): 47–60.

Li, Rita Yi Man 2017, forthcoming. Mart construction safety in road repairing works. *Procedia Computer Science* 103.

Li, Rita Yi Man, Simon Fong, and Hoi Man Chan. 2015. Cholars' comments and criticisms on construction and municipal solid waste management in Hong Kong and Singapore from 2000 to Present: A Review of Digital Information Online. In *Construction Safety and Waste Management: An Economic Analysis*, 31–45. Cham: Springer International Publishing.

Li, Rita Yi Man, Herru Ching Yu. Li, Cho Kei Mak, and Tang Beiqi Tang. 2016. Sustainable smart home and home automation: Big data analytics approach. *International Journal of Smart Home* 10 (8): 177–187.

Li, Rita Yi Man, and Sun Wah Poon, eds. 2013. *Supply of safety measures in developing and developed countries: A global perspective*. In *Construction Safety*, ed. Rita Yi Man Li and Sun Wah Poon. Germany: Springer.

Li, Rita Yi Man, Sun Wah Poon, and Daniel Chi Wing Ho. 2015. Content analysis on static digital information about the causes of construction accidents. In *Construction Safety and Waste Management: An Economic Analysis*, 47–60. Cham: Springer International Publishing.

Li, Rita Yi Man, and Sun Wah Poon. 2011. Using Web 2.0 to share the knowledge of construction safety as a public good in nature among researchers: The fable of economic animals. *Economic Affair* 31 (1): 73–79.

Li, Rita Yi Man, and Yi Lut Li. 2013. The role of competition law: An asian perspective. *Asian Social Science* 9 (7): 47–53.

Miorandi, Daniele, Sabrina Sicari, Francesco De Pellegrini, and Imrich Chlamtac. 2012. Internet of things: Vision, applications and research challenges. *Ad Hoc Networks no.* 10 (7): 1497–1516. doi:10.1016/j.adhoc.2012.02.016.

Perera, Charith, and Athanasios V. Vasilakos. 2016. A knowledge-based resource discovery for Internet of Things. *Knowledge-Based Systems no.* 109: 122–136. doi:10.1016/j.knosys.2016.06.030.

Podgórski, Daniel, Katarzyna Majchrzycka, Anna Dąbrowska, Grzegorz Gralewicz, and Małgorzata Okrasa. 2017. Towards a conceptual framework of OSH risk management in smart working environments based on smart PPE, ambient intelligence and the Internet of Things technologies. *International Journal of Occupational Safety and Ergonomics no.* 23 (1): 1–20.

Quest. *Waiting for the Internet of Things? IoT is already here.* 2017, cited 25 April 2017. Available from https://www.quest.com/ema-internet-of-things-infographic.

Teimourikia, Mahsa, and Mariagrazia Fugini. 2017. Ontology development for run-time safety management methodology in Smart Work Environments using ambient knowledge. *Future Generation Computer Systems*. doi:10.1016/j.future.2016.07.003.

Zhao, Jingzhu, Xiancao Zheng, Rencai Dong, and Guofan Shao. 2013. The planning, construction, and management toward sustainable cities in China needs the environmental internet of things. *International Journal of Sustainable Development & World Ecology no.* 20 (3): 195–198.

Chapter 8
RAND Appropriateness Study in Regard to Automated Construction Safety: A Global Perspective

Abstract The RAND appropriateness methodology was first used in medical research; the researchers aimed to compare and contrast different types of surgical methods based on various concerns in regard to these procedures. Assessors had to rate each procedure with a rating of from 1 to 9, where 1 indicated that the hazards of the procedure outweighed the benefits, and 9 signified that the benefits of the procedure outweighed the hazards. Since this beginning, many different variations of RAND have been developed and applied according to the requirements of various fields of research. In our research study, we first distributed a form (both online and on paper) which presented a 1 to 9 points RAND appropriateness survey. This was to look at the appropriateness of adopting different types of automated construction safety tools with regard to the following considerations: (1) construction safety; and (2) costs and benefits. The viewpoints of construction practitioners, academics and automated tool providers were canvassed from Pakistan, Hong Kong, Iran, Australia and mainland China. After that, we sent a "second-round" survey which provided the participants with the medians of the results from the first round. This study was the first of its kind—in terms of it adopting the medical RAND appropriateness methodology to construction research.

Keywords RAND appropriateness · Automated construction safety · Cost and benefits analysis

1 Introduction

The construction industry has long been criticised for its notoriously poor safety record. Despite the fact that many of these criticisms focus on the characteristics of workers that directly lead to accidents, such as age, experience, education and training (He et al. 2016; Li and Poon 2013), another strand of literature suggests that deficiencies in tools are among the major causes of accidents. Various automated tools have been developed which enhance safety on-site as well as facilitate cost reduction and efficiency improvements. For example, Hamledari et al. (2017)

R.Y.M. Li, *An Economic Analysis on Automated Construction Safety*,
DOI 10.1007/978-981-10-5771-7_8

introduced a computer vision-based algorithm which detects the components of an interior partition and infers its current state by using two-dimensional digital images. Also, Building Information Modelling (BIM) has become ubiquitous in recent years, whereby the project data and the building design are recorded and then manipulated using a digital format. BIM systems are frequently used in the construction stage and some of them can be used across the building's entire life cycle. In addition, drones have been used on sites for safety inspection (Kazemian et al. 2017).

Virtual reality technology has been applied to the manipulating of the cantilever in the construction of bridge decks; the technology has been used to assist in the planning stages of the incremental launching method for bridge deck construction. The interactive applications developed show the evolution of the construction work, monitor the planned construction processes, and visualise the details of every single one of the construction components (Sampaio and Martins 2014). Zhong et al. (2017) introduced a multidimensional IoT-enabled BIM platform to achieve real-time traceability and visibility in prefabricated public housing construction projects in Hong Kong. The project involved the building of five, 34–38 storey, residential buildings offering approximately 5000 public units for around 14,000 people.

2 Appropriateness

Appropriateness methodologies are often used in medical research. For example, Drumond et al. (2017) suggest that appropriateness can be used in relation to a set of pharmaceutical design features of a pharmaceutical product, and that it can determine the target patients and/or how caregivers can use the product. Kovacs et al. (2013) demonstrated the minimal percentage of lumbar spine magnetic resonance imaging sessions that are incongruously prescribed. In a study conducted by Cher et al. (2017), appropriateness criteria for drugs used for the active surveillance of prostate cancer show that "Article Low volume Gleason 6" is highly appropriate for active surveillance.

The idea of appropriateness has also been used in other areas of research although it is substantially less popular outside of the medical arena. For example, Lyons and Brockman (2017) shed light on the contribution of the "dark triad" and emotional expressivity to appropriateness of emotional response after exposure to stimuli. Tan et al. (2016) investigated food appropriateness. Unusual and novel foods such as insects are generally not considered to be appropriate as food, in principles; and they are not often eaten for reasons other than their intrinsic sensory properties, for example, ant. That study explored the way in which the levels of sensory-liking and food appropriateness affect the likelihood that unusual foods will be eaten.

3 Institutional Theory

Institutional theory has been applied in various fields such as political science, organisational studies and economics (Wang et al. 2012). The theory has also been applied to housing-related topics (Li 2011, 2014) and construction safety research (Li and Poon 2011). Under the lens of economics, it can be said that institutions represent the rules of games that shape human interactions. In general, there are two types of institutions: formal and informal. The former ones generally refer to laws which are written down (Tamanaha 2015). Institutional norms, i.e. informal and unwritten rules (Wang et al. 2012), always play an important role in employees' behaviours, and organisations' decisions are driven by social and cultural factors (Boateng and Agyemang 2015; Wang et al. 2014). Different norms and perceptions are displayed by different groups due to differences in occupation, education and organisational rationalities with regard to changes and problem solving.

Institutions are considered as a behavioural reference which individuals are urged to follow (Wang et al. 2012), a scaffolds of constraints on stakeholders, and something which contributes to the process of variation across different times. To a certain extent, the institutional structures which are inherited from the past are the root that results in the institution being impervious to change, as the proposed changes are counter to previous beliefs or threaten the leaders of the existing organisation (Von Staden and Bruce 2015).

4 Cost/Benefit Analysis

A prudent enterprise undertakes a cost–benefit analysis whenever it makes decisions with regard to whether to deploy new systems and services. Some costs and benefits can be measured qualitatively, such as customer and employee satisfaction and the stress placed on staff by the change, whereas others are measured quantitatively based on quantities of materials, time, money, etc. Some costs and benefits can be measured directly and others cannot: for instance, the impact of a lost business opportunity resulting from potential customers not being provided with timely and correct information (Wu et al. 2013).

Wu et al. (2013)'s research suggests that there are different approaches which are adopted in cost and benefit analysis. For example, the subjective analysis approach asks decision-makers to subjectively determine whether the prospective benefits of, for instance, an information system are worth the projected costs; this approach is used when the benefits are intangible or uncertain. On the other hand, the cost-effectiveness approach is used for choosing among similar information systems or system components, but does not explicitly address whether the benefit of any of the systems exceeds their costs.

Vast amounts of the research in the literature utilise the cost–benefit approach. For example, Chen employs a cost–benefit analysis to evaluate the economic efficiency of car recycling policies (Chen et al. 2015). Li (2010) suggests that collaboration between countries on environmental issues depends on the cost and benefit for each of the country.

5 The RAND Appropriateness Research Method

The RAND appropriateness method was initially used in medical research to study the appropriateness of different procedures (mainly surgical ones). In such a study, a number of panellists have to rate the appropriateness of each method according to the risks and benefits of using these approaches. Such a study essentially combines the best available scientific evidence with the collective judgment of experts. The RAND appropriateness methodology originally aimed to yield a result regarding the appropriateness of performing a medical procedure (Jones et al. 2016). It is a statistically validated methodology that incorporates expert consensus, and this methodology has been shown to produce appropriateness criteria which demonstrate construct and predictive ability (Brar et al. 2013).

It can be seen from previous research that different researchers have used different forms of the RAND appropriateness methodology. For example, Jones et al. (2016)'s initial first-round rating was conducted remotely via an online survey while Cher et al. (2017)'s initial round provided each of the panellist a ratings score sheet, definitions and instructions along with a PowerPoint presentation.

In the second round, Cher et al. (2017)'s scoring was conducted at a face-to-face meeting. Each panellist was provided with an individualised document showing his or her first-round ratings along with the statistical distribution of first-round ratings from all the other panellists. In other words, for every scenario, every panel member could see where his or her scores stood relative to the scores of all the other panellists. Scenarios were discussed and debate was encouraged regarding the scenarios with wider dispersions of scores (i.e. those demonstrating less agreement) (Cher et al. 2017).

The panel chair moderated the discussion with assistance from the expert on the method. During the discussion, minor changes were made to the counselling statements in order to produce the final versions. Panellists understood that the median score and its associated terminology would be the primary end point for each scenario, and disagreement would be assessed by examining the standard deviations exhibited by the scores. Second-round scoring was completed at the meeting. In order to understand the degree to which patient-based parameters affected the final scores an ordinal regression model was used. An ordinal logistic regression model was applied to evaluate the results (Cher et al. 2017). On the other hand, Brar et al. (2013)'s form of RAND placed more stress on the discussion of the areas of disagreement (Fig. 1).

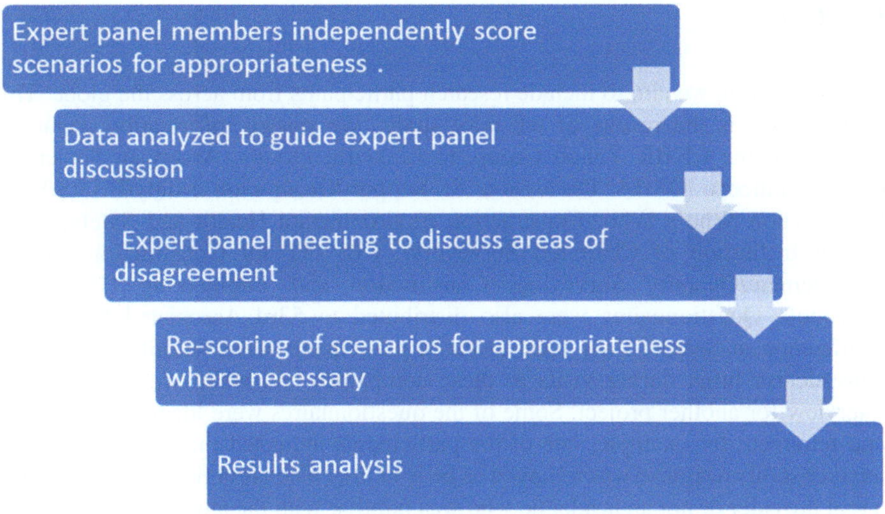

Fig. 1 Procedure adopted by Brar et al. (2013)

Nevertheless, no matter where the variations lay, the panellists in all these studies were instructed to use their best clinical judgment to rate each scenario on a scale of 1–9, where 1 indicated that the harms of AS outweighed the benefits, and 9 signified that the benefits of AS outweighed the problems (Cher et al. 2017).

The scores were associated with the RAM terminology and ranged from highly inappropriate (score 1) to highly appropriate (score 8–9); they were also associated with paradigmatic counselling statements (Cher et al. 2017): the benefit-to-harm ratio of a procedure on a scale of 1–9, where 9 means that the expected benefits significantly prevail over the expected harms, and 1 means that the expected harms prominently greater than the expected benefits. A middle rating of 5 means either that the harms and benefits are roughly equal or that the rater cannot make any judgement for the patient (Saccullo et al. 2013) (Table 1).

Table 1 RAND appropriateness values representation (Cher et al. 2017; Jones et al. 2016)

Score	RAND/UCLA term (Cher et al. 2017)	Median value	RAND interpretation (Jones et al. 2016)
1	Highly inappropriate	1–3	Inappropriate (without disagreement)
2–3	Inappropriate		
4–5	Uncertain	4–6	Uncertain
6–7	Appropriate	7–9	Appropriate (without disagreement)
8–9	Highly appropriate		

6 Results

In this research, we aim to include research participants from across the globe. This chapter records the results of the pilot study. We posted information about the research to the CNBR Yahoo Group, to LinkedIn groups (American Society of Civil Engineers, OSHA Discussion & Support/Occupation safety/EH&S/HSE, Construction Who's Who, Construction Environmental Health and Safety, RICS (Royal Institution of Chartered Surveyors), CIB—IT for Construction, Project Management/Quantity Surveying, Construction and Engineering Group and Facebook. Questionnaires were also distributed to 53rd Associated Schools of Conference in Seattle. Four local research participants were recruited from local construction firms during visits to these companies and when we conducted an interview for another project. Some of the questionnaires were identified as invalid and removed; for example, one of the participants indicated that he is an English lecturer so his response was removed as he is neither a construction practitioner, nor a tool provider or nor an academic who was familiar with the automated tools under review. Other responses were removed as being incomplete.

Fifteen research participants were recruited in the first round, in total. About half of the research participants worked for companies with one hundred or fewer employees. Most of them have over 6 years of work experience. Over half, 53% of participants, came from the construction industry, 33.3% came from the construction industry and 13.3% were tool providers. Nearly 70%, that is 66.7% of the participants were male, but that, in fact, reflects the reality that most construction practitioners, associated academics and tool providers are male; it is a male-dominated industry. One third, 33.3%, of participants had undergraduate degrees, and 66.7% had postgraduate qualifications. Some 40% of them reported that they would break down problems and solve them step-by-step while the remaining ones were reactive problem solvers. Approximately one-quarter of the respondents came from Hong Kong, others came from the UK (Oxford and another city), the United States (Las Vegas and Lexington), Australia (Melbourne), Pakistan (Mardan) and mainland China (Fujian, Shenzhen and Guangdong).

Most of the research participants reported doubt about the appropriateness of using 3D printing to print an entire building (this attained the lowest score among all the four proposals). A median value of 7 was recorded (the highest score among all the proposals) for the proposals that: (1) VR should be used for safety training to reduce construction accidents; (2) robots could be used for some dangerous jobs and this would reduce construction accidents; (3) big data could be collected through IoT on sites; (4) IoT should be adopted to reduce construction accidents on sites; and (5) BIM should be used to better foresee structural building problems before the building is constructed.

Under the lens of economics, it can be seen that the cultures were different across the different groups. In this present study, we found that the academics believed that VR could convey safety knowledge with a median score of 7, but construction practitioners recorded a median score for this of only of 5.5. Academics considered using 3D printing to print a model of a building in order to visualise potential safety issues to be highly appropriate with a median score of 8 (indicating that the benefits exceeded the costs). However, construction practitioners only placed a median score of 5.5 on this. Academics considered that robots are highly appropriate for some dangerous jobs (this yielded a median score of 8) but the other two groups only scored this proposal at a median of 6.5. On the other hand, whilst all the groups placed a lower value on the proposal to use 3D printing to print entire buildings, the tool providers were relatively optimistic about this, though even their scoring was quite low (Figs. 2, 3, 4, 5, 6 and 7 and Tables 2 and 3).

For the second round, we sent an email to all those who had provided us with their email addresses in the first round. We also invited people who had responded via LinkedIn and CNBR to fill in the second-round questionnaires. Seven valid questionnaires were collected. Two were from construction practitioners, two came from academia and three were from tool providers. Overall, four second-round respondents were male and two were female. Over the past 5 years, these people had mainly worked in Seoul, Shenzhen, Surat, Guangdong, Lexington and Hong Kong. One indicated that he had worked in London, Washington and Oslo. Two of the organisations that the respondents had worked for were considered to be reactive to problems, two were considered to anticipate change and shape their environment actively, three of these organisations were said to break down problems and solve them step-by-step. Most of the results were found to be similar to those of the first round with the exception of the results pertaining to the use of robots; in regard to these, most of the answers had reduced in value (in terms of appropriateness). That is, in the second round the respondents considered that adopting the use of robots on sites to reduce construction accidents was less certain to be of value than they had considered this to be in the first round. The most appropriate proposals were now considered to be the ones involving adopting VR

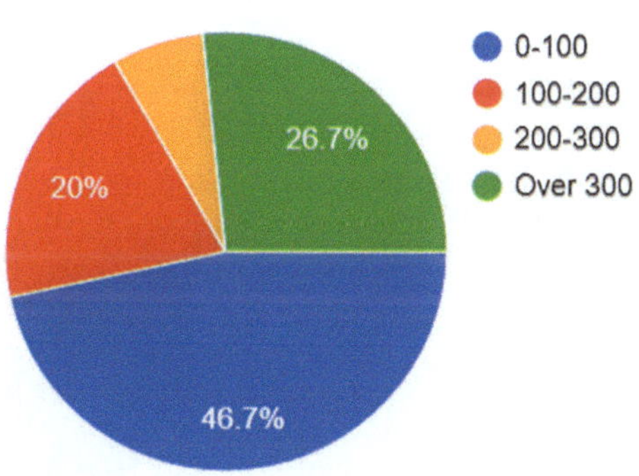

Fig. 2 Size of the company that the respondents worked for

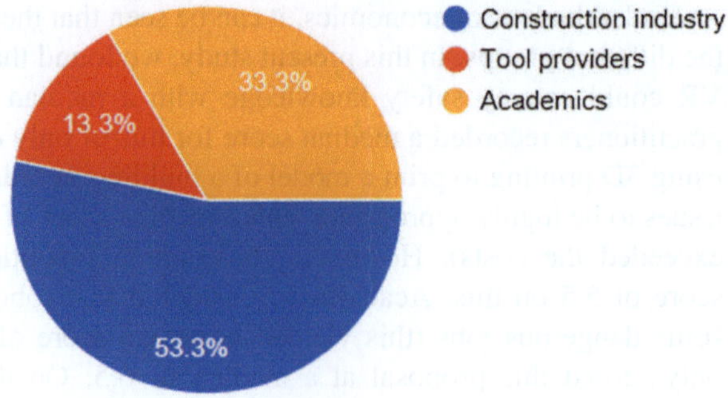

Fig. 3 Nature of the company that the respondents worked for

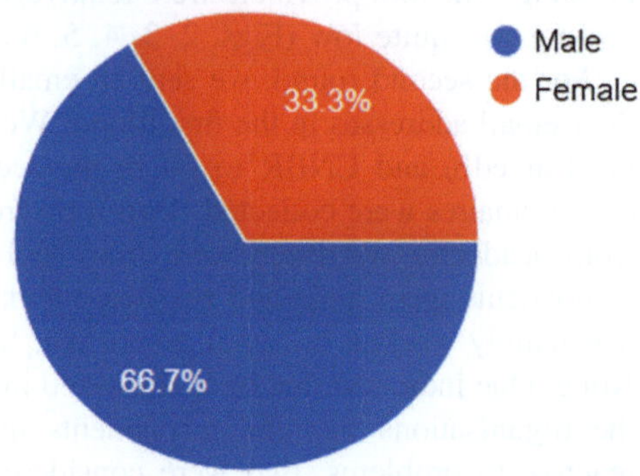

Fig. 4 Sex of the research participants

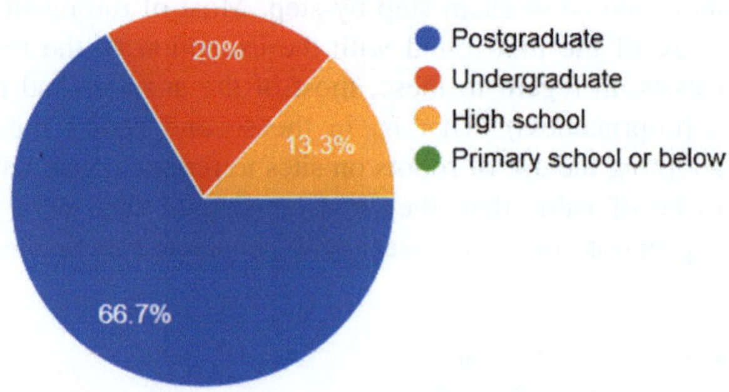

Fig. 5 Educational background of the respondents

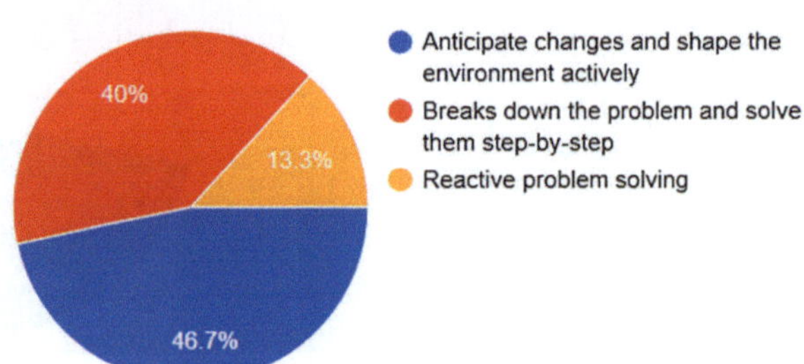

Fig. 6 Types of the institutions that the respondents worked for

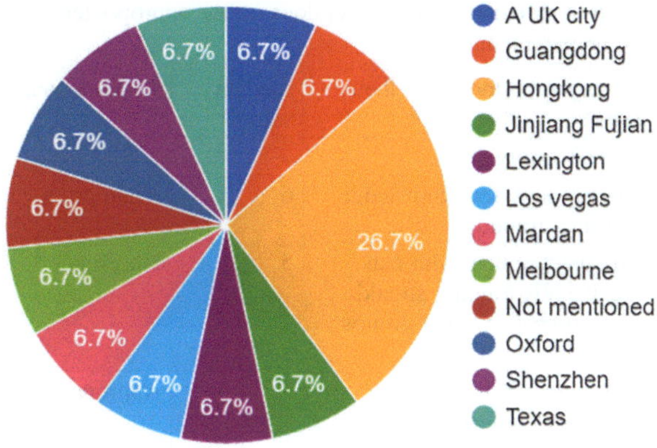

Fig. 7 Places that respondents had often worked over the past 5 years

for safety training to reduce construction accidents and the ones involving adopting IoT to reduce construction accidents on construction sites (Table 4).

7 Discussion and Conclusion

In this research, we used a modified RAND approach to study the appropriateness of adopting the various uses of automated tools proposed in order to reduce accidents on construction sites. This chapter presents the results of this pilot study. In general, academics are the group with the most optimistic view of the benefits that the automated tools would result in. Construction practitioners are quite pessimistic in this regard. The most appropriate proposals were considered to be those involving the adoption of VR for safety training to reduce construction accidents and those involving the adoption of IoT to reduce construction accidents on sites. These two types of proposal received the highest median appropriateness value.

One shortcoming of these research results is that the number of research participants was quite small as compared to the whole population of people involved in the construction industry across the world (including academics). Besides, as the research participants were scattered across many different parts of the world, face-to-face discussion on the proposals which resulted in the most disagreements was not included in this study.

Table 2 Median value, perceptions on the appropriateness and costs and benefits of the tools of all the participants

Name of the tools and their uses	Median value	Appropriateness	Costs and benefits interpretation
Virtual reality (VR)			
Conveying safety knowledge in safety training	6	Appropriate	The expected benefits outweigh the expected costs
The practitioners who can receive VR training should be restricted to some groups only	5	Uncertain	Either that the costs and benefits are about equal or that the rater cannot make a judgement
Adopting VR for safety training to reduce construction accidents	7	Appropriate	The expected benefits outweigh the expected costs
Additive manufacturing (3D printing)			
3D printing is used to print tan entire building	4	Uncertain	Either that the costs and benefits are about equal or that the rater cannot make the judgement
Limited range of materials should be used in 3D printing	5	Uncertain	Either that the costs and benefits are about equal or that the rater cannot make the judgement
3D printing is used to print a model of a building in order to visualise potential safety issues	6	Appropriate	The expected benefits outweigh the expected costs
Robot			
As labour costs are skyrocketing, robots should replace labour	5	Uncertain	Either that the costs and benefits are about equal or that the rater cannot make the judgement
Wearable robotics can provide strength to workers, reduces the likelihood of fatigue	6	Appropriate	The expected benefits outweigh the expected costs
Robots should be used for some dangerous jobs	7	Appropriate	The expected benefits outweigh the expected costs
Adopting robots on sites to reduce construction accidents	7	Appropriate	The expected benefits outweigh the expected costs
Internet of Things (IoT)			
To store the positions/routes of the workers	6	Appropriate	The expected benefits outweigh the expected costs
Big data collected through IoT on sites	7	Appropriate	The expected benefits outweigh the expected costs

(continued)

Table 2 (continued)

Name of the tools and their uses	Median value	Appropriateness	Costs and benefits interpretation
Adopting IoT to reduce construction accidents on sites	7	Appropriate	The expected benefits outweigh the expected costs
Building Information Modelling (BIM)			
BIM visualises real-time status	6	Appropriate	The expected benefits outweigh the expected costs
Using BIM to better foresee a building's structural problems before the building is constructed	7	Appropriate	The expected benefits outweigh the expected costs
Adopting BIM to reduce construction accidents	6	Appropriate	The expected benefits outweigh the expected costs

Appendix 1

Questions Listed in the RAND

In this study, we aim to look at construction/tools providers/academics' viewpoint with regards to the advanced automated tools (virtual reality, additive manufacturing, robots, the Internet of Things and BIM) which can be applied in the construction industry. There is no definite right or wrong answer to the questions posed. We will analyse the data in an aggregate manner and individual identities are not revealed. Should you have any enquiries, you may send an email to the principal researcher ymli@hksyu.edu, Hong Kong Shue Yan University.

In the following question, we wish to know your viewpoint on the automated tool with regard to **construction safety**:

Table 3 Impact of occupation background, education and organisation characteristics on the appropriateness/costs and benefits in using the tools for enhancing construction site safety

Name of the tools and their uses	Academics (n = 5)	Construction practitioners (n = 8)	Tool providers (n = 2)	Undergraduate or below	Postgraduate	Anticipate changes and shape the environment actively	Breaks down the problem and solve them step-by-step	Reactive problem solving
Virtual reality (VR)								
Conveying safety knowledge in safety training	7	5.5	6.5	7	6	6	6.5	7
Number of the practitioners who can receive VR training is restricted in some groups only	6	4.5	5.5	5	5.5	5	5	6.5
Adopting VR for safety training to reduce construction accidents	7	5.5	7	7	6.5	6	6.5	7.5
3D printing								
3D printing is used to print the whole building out	4	4	5.5	4	5	4	4	5
Limited range of materials can be used in 3D printing	7	5	6	4	5.5	5	5	6.5

(continued)

Table 3 (continued)

Name of the tools and their uses	Academics (n = 5)	Construction practitioners (n = 8)	Tool providers (n = 2)	Undergraduate or below	Postgraduate	Anticipate changes and shape the environment actively	Breaks down the problem and solve them step-by-step	Reactive problem solving
3D printing is used to print the building model out to visualise potential safety issue	8	5.5	6.5	7	6	4	6.5	7.5
Robot								
As labour cost is skyrocketing, robots can replace labour	6	4	5.5	5	5	5	3.5	5
Wearable robotics provides strength to workers, reduces the likelihood of fatigue	6	6.5	5.5	6	6.5	6	6	6
Robots are used for some dangerous jobs	8	6.5	6.5	5	7.5	8	5	8
Adopting robots on sites to reduce construction accidents	6	7.5	6	6	7	8	5.5	6

(continued)

Table 3 (continued)

Name of the tools and their uses	Academics (n = 5)	Construction practitioners (n = 8)	Tool providers (n = 2)	Undergraduate or below	Postgraduate	Anticipate changes and shape the environment actively	Breaks down the problem and solve them step-by-step	Reactive problem solving
Internet of Things (IoT)								
It saves the positions/routes of the workers	5	6	6	4	6	6	5.5	6.5
Big data collected through IoT on sites	7	7	6.5	5	7	7	6.5	7
Adopting IoT to reduce construction accidents on sites	4	7	7	5	7	6	6	7
Building Information Modelling (BIM)								
BIM visualises real-time status	5	6.5	6	5	7	7	5	7
Better foresee building structure problem before the building is constructed	7	6.5	6.5	6	7	7	6	7
Adopting BIM to reduce construction accidents	6	7	5.5	5	7	8	5	6.5

Table 4 Results of second round after median values of first round are known

Name of the tools and their uses	Median value	Median
Virtual reality (VR)		
Conveying safety knowledge in safety training	No change	6
The practitioners who will receive VR training should be restricted to some groups only	−1	4
Adopting VR for safety training to reduce construction accidents	No change	7
Additive manufacturing (3D printing)		
3D printing should be used to print entire buildings	No change	4
Limited range of materials can be used in 3D printing	+1	6
3D printing should used to print a model of a building model out in order to visualise potential safety issues	−2	4
Robot		
As labour costs are skyrocketing, robots should replace labour	No change	5
Wearable robotics can provides extra strength to workers, reducing the likelihood of fatigue	−1	5
Robots can be used for some dangerous jobs	−2	5
robots can be used on sites to reduce construction accidents	−3	4
Internet of Things (IoT)		
IoT should store the positions/routes of the workers	No change	6
Big data should be collected through IoT on sites	−2	5
Adopting IoT to reduce construction accidents on sites	No change	7
Building Information Modelling (BIM)		
BIM visualises real-time status	No change	6
BIM can be used to better foresee building structural problems before the building is constructed	−1	6
Adopting BIM to reduce construction accidents	−1	5

Score	Meaning of the score	Benefit-to-cost ratio of the tools
1	Highly inappropriate	The expected costs greatly outweigh the expected benefits
2–3	Inappropriate	Between 1 and 4
4–5	Uncertain	Either that the costs and benefits are about equal or that the rater cannot make the judgement
6–7	Appropriate	Between 5 and 8
8–9	Highly appropriate	The expected benefits greatly outweigh the expected costs

Virtual reality (VR)	Score
Conveying safety knowledge in safety training	
Number of the practitioners who can receive VR training is restricted in some groups only	
Adopting VR for safety training to reduce construction accidents	
3D printing	
3D printing is used to print the whole building out	
Limited range of materials can be used in 3D printing	
3D printing is used to print the building model out to visualise potential safety issue	
Robots	
As labour cost is skyrocketing, robots can replace labour	
Wearable robotics provides strength to workers, reduces the likelihood of fatigue	
Robots are used for some dangerous jobs	
Adopting robots on sites to reduce construction accidents	
Internet of Things (IoT)	
It sends a warning to the worker when he or she is inactive for a long period	
It saves the position/route of workers	
Adopting IoT to reduce construction accidents on sites	
Building Information Modelling (BIM)	
BIM visualises real-time status	
Better foresee building structure problem before the building is constructed	
Adopting BIM to reduce construction accidents	

Background information

Age	
Years of experience in construction industry	
Size of the company 0–100 100–200 200–300 Over 300	
Occupation	
Name of the company	
Nature of the company Construction industry Tool providers Academics	
Female Male	
Education Postgraduate Undergraduate High school Primary school or below	

(continued)

(continued)

Age	
Organisation type Anticipate changes and shape the environment actively/ Breaks down the problem and solve them step-by-step/ Reactive problem solving	
Major city that you work over the past 5 years	
The following question is required for construction industry practitioners only	
Has your company adopted the following tools?	
Robot Yes No	
Addictive manufacturing Yes No	
Internet of Things Yes No	
Virtual Reality Yes No	
Building Information Modelling Yes No	

Note the median values for each of the questions in page 1 are given

References

Boateng, H., and F.G. Agyemang. 2015. The rise of culture in knowledge sharing in a public-sector organisation in Ghana: Revisiting Hofstede's Model. *International Journal of Public Administration* 38 (7): 486–495.

Brar S., C. Law, R. McLeod, L. Helyer, C. Swallow, L. Paszat, R. Seevaratnam, R. Cardoso, M. Dixon, A. Mahar, L.G. Lourenco, L. Yohanathan, A. Bocicariu, T. Bekaii-Saab, I. Chau, N. Church, D. Coit, C.H. Crane, C. Earle, P. Mansfield, N. Marcon, T. Miner, S.H. Noh, G. Porter, M.C. Posner, V. Prachand, T. Sano, C. van de Velde, S. Wong, S. and N. Coburn 2013. Defining surgical quality in gastric cancer: A RAND/UCLA Appropriateness Study. *Journal of the American College of Surgeons*, 217(2): 347–357.e341.

Chen, Z., D. Chen, T. Wang, and S. Hu. 2015. Policies on end-of-life passenger cars in China: dynamic modeling and cost-benefit analysis. *Journal of Cleaner Production* 108: 1140–1148.

Cher, M.L., A. Dhir, G.B. Auffenberg, S. Linsell, Y. Gao, B. Rosenberg, S.M. Jafri, L. Klotz, D.C. Miller, K.R. Ghani, S.J. Bernstein, J.E. Montie, and B.R. Lane. 2017. Appropriateness criteria for active surveillance of prostate cancer. *The Journal of Urology* 197 (1): 67–74.

Drumond, N., D.A. van Riet-Nales, F. Karapinar-Çarkit, and S. Stegemann. 2017. Patients' appropriateness, acceptability, usability and preferences for pharmaceutical preparations: Results from a literature review on clinical evidence. *International Journal of Pharmaceutics* 521 (1–2): 294–305.

Hamledari, H., B. McCabe, and S. Davari. 2017. Automated computer vision-based detection of components of under-construction indoor partitions. *Automation in Construction* 74: 78–94.

He, Q., S. Dong, T. Rose, H. Li, Q. Yin, and D. Cao. 2016. Systematic impact of institutional pressures on safety climate in the construction industry. *Accident Analysis and Prevention* 93: 230–239.

Jones, E., S. Beriwal, D. Beyer, J. Chino, A. Jhingran, L. Lee, J. Michalski, A.J. Mundt, C. Patton, I. Petersen, L. Portelance, J.K. Schwarz, and S. McCloskey. 2016. An analysis of appropriate delivery of postoperative radiation therapy for endometrial cancer using the RAND/UCLA Appropriateness Method. *Advances in Radiation Oncology* 1 (1): 26–34.

Kazemian, A., X. Yuan, E. Cochran, and B. Khoshnevis. 2017. Cementitious materials for construction-scale 3D printing: Laboratory testing of fresh printing mixture. *Construction and Building Materials* 145: 639–647.

Kovacs, F.M., E. Arana, A. Royuela, A. Cabrera, C. Casillas, P. Piñero, M. Vega, B. Asenjo, A. Estremera, G. Amengual, H. Sarasíbar, P. Ferrer, A. Manjarrés, J. Zamarro, F. Bravo-Rodríguez, J.C. Paniagua, J. Mota, T. Sánchez-Sagrado, V. Abraira, and C. Martínez. 2013. Appropriateness of lumbar spine magnetic resonance imaging in Spain. *European Journal of Radiology* 82 (6): 1008–1014.

Li, R.Y.M. 2010. Constructing sustainable vertical cities: Strategies to enhance closer cooperation between ASEAN contractors on pollution problem under the Lens of economic game theories, cost benefit analysis and dynamic markov chain theories. *Asian Social Science* 6 (5): 10–18.

Li, R.Y.M. 2011a. *Everyday life application of Neo-institutional Economics*. Saarbrucken: Lambert Academic Publishing.

Li, R.Y.M. 2014. *Law, economics and finance of the real estate market: A perspective of Hong Kong and Singapore*. Berlin: Springer.

Li, R.Y.M., and Poon, S.W. 2011. Using Web 2.0 to Share the knowledge of construction safety as a public good in nature among researchers: the fable of economic animals. *Economic Affair* 31 (1): 73–79.

Li, R.Y.M., and S.W. Poon. 2013. A literature review on the causes of construction accidents. In *Construction Safety*, ed. R.Y.M. Li, and S.W. Poon. Germany: Springer.

Lyons, M., and C. Brockman. 2017. The Dark Triad, emotional expressivity and appropriateness of emotional response: Fear and sadness when one should be happy? *Personality and Individual Differences* 104: 466–469.

Saccullo, G., M. Marietta, M. Carpenedo, V. De Stefano, A. Falanga, A.B. Federici, F. Rodeghiero, A. Tosetto, and S. Siragusa. 2013. Platelet cut-off for anticoagulant therapy in cancer patients with venous Thromboembolism and Thrombocytopenia: An expert opinion based on RAND/UCLA appropriateness method (RAM). *Blood* 122 (21): 581.

Sampaio, A.Z., and O.P. Martins. 2014. The application of virtual reality technology in the construction of bridge: The cantilever and incremental launching methods. *Automation in Construction* 37: 58–67.

Tamanaha, B.Z. 2015. The knowledge and policy limits of new institutional economocs on development. *Journal of Economic Issues* 49 (1): 89–109.

Tan, H.S.G., A.R.H. Fischer, H.C.M. van Trijp, and M. Stieger 2016. Tasty but nasty? Exploring the role of sensory-liking and food appropriateness in the willingness to eat unusual novel foods like insects. *Food Quality and Preference* 48(Part A): 293–302.

Von Staden, P., and K. Bruce. 2015. Original and new institutional economics: Brethren rather than foes? Lessons from the sociocognitive turn in "Late" Douglass North. *Journal of Economic Issues* 49 (1): 111–125.

Wang, H.K., J.F. Tseng, and Y.F. Yen. 2012. Examining the mechanisms linking Guangxi, norms and knowledge sharing: The mediating roles of trust in Taiwan's high-tech Firms. *The International Journal of Human Resource Management* 23 (19): 4048–4068.

Wang, H.K., J.F. Tseng, and Y.F. Yen. 2014. How do institutional norms and trust influence knowledge sharing? An institutional theory. *Innovation Management, Policy and Practice* 16 (3): 374–391.

Wu, M., A. Turpin, J. Thom, F. Scholer, and R. Wilkinson. 2013. Cost and benefit estimation of experts' mediation in an enterprise search. *Journal of the Association for Information Science and Technology* 65 (1): 146–163.

Zhong, R.Y., Y. Peng, F. Xue, J. Fang, W. Zou, H. Luo, S. Thomas Ng, W. Lu, G.Q.P. Shen, and G.Q. Huang. 2017. Prefabricated construction enabled by the Internet-of-Things. *Automation in Construction* 76: 59–70.

MIX
Papier aus verantwortungsvollen Quellen
Paper from responsible sources
FSC® C105338

If you have any concerns about our products,
you can contact us on
ProductSafety@springernature.com

In case Publisher is established outside the EU,
the EU authorized representative is:
Springer Nature Customer Service Center GmbH
Europaplatz 3, 69115 Heidelberg, Germany

Printed by Libri Plureos GmbH
in Hamburg, Germany